"SCIENCE BOOK OF THE YEAR"
—*The Times* (London)

A BEST BOOK OF THE YEAR
Smithsonian, Science Friday, Popular Mechanics, Science News, Library Journal, Booklist, Chicago Public Library

A *New York Times, Globe and Mail* (Toronto), *Sunday Times* (London) BESTSELLER

GOODREADS CHOICE AWARDS WINNER

"*The Rise and Fall of the Dinosaurs* is a triumph, written by one of our young leaders of the field."
—Neil Shubin, author of *Your Inner Fish*

"Scintillating. . . . Brusatte's mastery of his field, formidable explanatory powers and engaging style have combined to produce a masterpiece of science writing." —*Washington Post*

"If John McPhee's love affair with rocks in *Annals of the Former World* floats your boat . . . you're going to love *The Rise and Fall of the Dinosaurs*. . . . [The] emotional connection, and Brusatte's collection of personal stories and characters, make his book special." —*New York Times Book Review*

"An absorbing historical saga. . . . Skillfully combines interesting and amusing stories from his field experience with the broader story of dinosaurs ruling the Earth. . . . Brusatte's recounting of the millions of years of dinosaur dominance makes for very nerdy, very thrilling reading." —*Christian Science Monitor*

"Ripping yarns from the age of dino might. . . . *The Rise and Fall of the Dinosaurs* is a lovely book. Brusatte has a wonderful knack for conjuring vivid worlds out of a few shards of petrified bone. He is excellent company as a narrator. . . . This is a fine piece of writing that drags the dinosaurs out of museums and reanimates them for a new generation."
—*The Times* (London)

"Fascinating. . . . This is popular-science writing at its best."

—*Booklist* (starred review)

"Captivating. . . . First-rate science writing. . . . Superb."

—*Publishers Weekly* (starred review)

"Full of adventures and humor. . . . Readers will come away from this book with a greater appreciation for the great strides dinosaur paleontologists have made in the past few decades."

—*Science*

"Brusatte does for dinosaurs what E.O. Wilson did for ants and Carl Sagan for stars. . . . If you ever loved a dinosaur, buy this book."

—*Washington Times*

"Brusatte delivers a cutting-edge account of Earth's most awe-inspiring age—and does so with great skill, humor, and wonder. . . . [A] thrilling account . . . told vividly by one of the world's top paleontologists."

—Peter Brannen, author of *The Ends of the World*

"*The Rise and Fall of the Dinosaurs* is an expansive biography of this peerless group of species, intermixed with insights about the methods paleontologists use to reconstruct their lost world."

—VICE/Motherboard

"Vibrant. . . . [Brusatte] is as adept a scientific storyteller as any reader could ask for."

—*The Spectator* (UK)

"[A] Jurassic blockbuster. . . . A gripping read in the best traditions of popular science. . . . Infectiously ebullient."

—*The Observer* (London)

"*The Rise and Fall of the Dinosaurs* may be the best book on dinosaurs ever written for a popular audience."

—Inside Higher Ed

"Vivid. . . . This book will change the way you think about dinosaurs."

—Gizmodo

The **RISE** *and* **FALL** *of the* **DINOSAURS**

The RISE *and* FALL *of the* DINOSAURS

A New History of Their Lost World

STEVE BRUSATTE

wm

WILLIAM MORROW

An Imprint of HarperCollins*Publishers*

For Mr. Jakupcak, my first and finest teacher of paleontology, and my wife, Anne, and all others teaching the next generation.

HarperCollins books may be purchased for educational, business, or sales promotional use. For information, please email the Special Markets Department at SPsales@harpercollins.com.

A hardcover edition of this book was published in 2018 by William Morrow, an imprint of HarperCollins Publishers

FIRST WILLIAM MORROW PAPERBACK EDITION PUBLISHED 2019.

Designed by Bonni Leon-Berman

Chapter title art by Todd Marshall
World Maps of the Prehistoric Earth © 2016 Colorado Plateau Geosystems, Inc.

Where no credit is specified, photograph is courtesy of the author.
Additional photo credit information: Page 111, first image: Image #36246a, American Museum of Natural History Library. Page 131: Image #238372, American Museum of Natural History Library. Page 132, first image: Image #328221, American Museum of Natural History Library. Page 132, second image: Image #312963, American Museum of Natural History Library. Page 133: Published in Maidment et al., PLoS ONE, 2015, 10 (10): e0138352. Page 169: Image #17808, American Museum of Natural History Library. Page 198–9: Image #00005493, American Museum of Natural History Library.

The Library of Congress has catalogued a previous edition of the book as follows:

Names: Brusatte, Stephen, author.
Title: The rise and fall of the dinosaurs : a new history of a lost world / Stephen Brusatte.
Description: New York, NY : William Morrow, [2018] | Includes bibliographical references and index.
Identifiers: LCCN 2017038066 | ISBN 9780062490421 (hardcover)
Subjects: LCSH: Dinosaurs.
Classification: LCC QE861.4 .B79 2018 | DDC 567.9—dc23
LC record available at https://lccn.loc.gov/2017038066

ISBN 978-0-06-249043-8 (pbk.)

20 21 22 23 10 9 8 7

CONTENTS

TIMELINE OF THE AGE OF DINOSAURS

PALEOZOIC ERA	MESOZOIC ERA								CENOZOIC ERA	
Permian	***Triassic***			***Jurassic***			***Cretaceous***		***Paleogene***	***Period***
	Early	Middle	Late	Early	Middle	Late	Early	Late		Epoch
	252-247	247-237	237-201	201-174	174-164	164-145	145-100	100-66		Age (Millions of years ago)

DINOSAUR FAMILY TREE

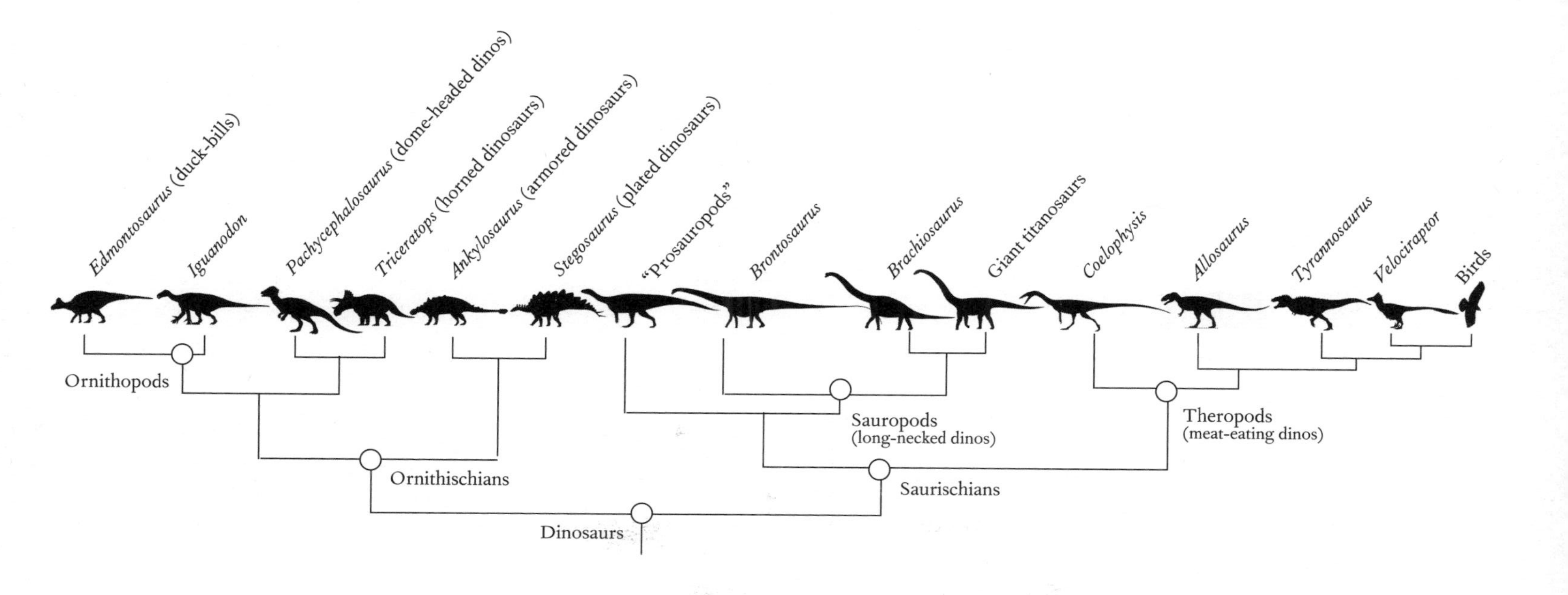

WORLD MAPS OF THE PREHISTORIC EARTH

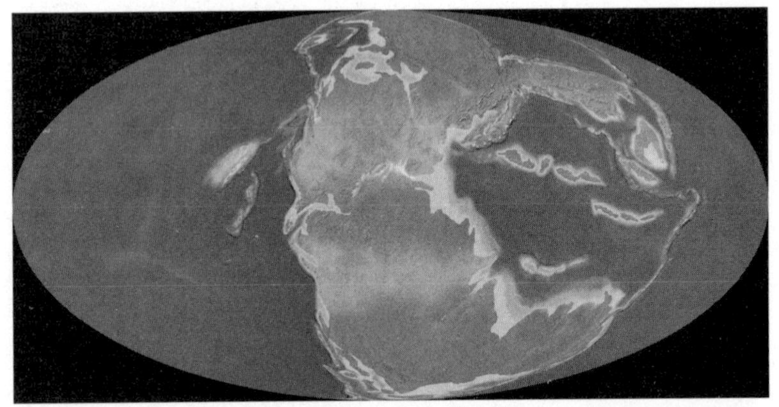

Triassic Period (ca. 220 million years ago)

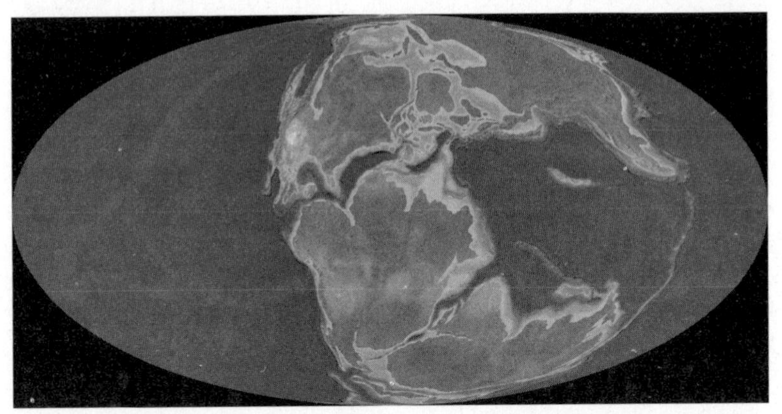

Late Jurassic Period (ca. 150 million years ago)

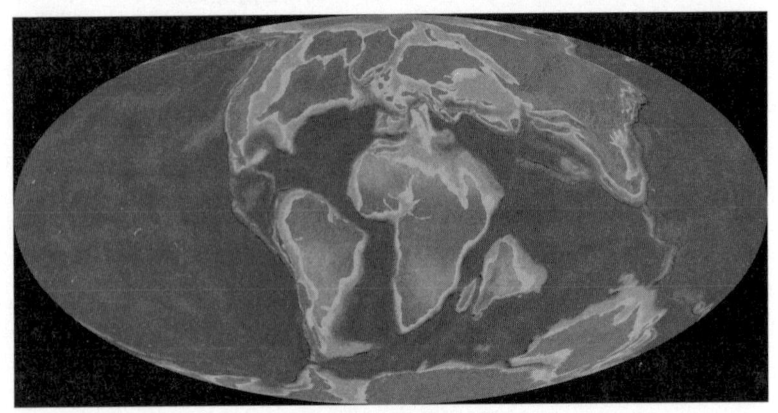

Late Cretaceous Period (ca. 80 million years ago)

The **RISE** *and* **FALL** *of the* **DINOSAURS**

PROLOGUE:

The GOLDEN AGE *of* DISCOVERY

Zhenyuanlong

A FEW HOURS BEFORE LIGHT broke on a cold November morning in 2014, I got out of a taxi and pushed my way into Beijing's central railway station. I clutched my ticket as I fought through a swarm of thousands of early-morning commuters, my nerves starting to jangle as the departure time for my train ticked ever closer. I had no idea where to go. Alone, with only a few words of Chinese in my vocabulary, all I could do was try to match the pictographic characters on my ticket to the symbols on the platforms. Tunnel vision set in, and I barreled up and down the escalators, past the newsstands and noodle joints, like a predator on the hunt. My suitcase—weighed down with cameras, a tripod, and other scientific gear—bounced along behind me, rolling over feet and smacking into shins. Angry shouts seemed to come at me from all directions. But I didn't stop.

By now sweat was pouring through my downy winter jacket, and I was gasping in the diesel haze. An engine roared to life somewhere ahead of me, and a whistle sounded. A train was about to depart. I staggered down the concrete steps leading to the tracks and, to my great relief, recognized the symbols. Finally. This was my train—the one that would be shooting northeastward to Jinzhou, a Chicago-size city in old Manchuria, a few hundred miles from the frontier with North Korea.

For the next four hours, I tried to get comfortable as we crawled past concrete factories and hazy cornfields. Occasionally I nodded off, but I couldn't steal back much sleep. I was far too excited. A mystery was waiting at the end of the journey—a fossil that a farmer stumbled upon while harvesting his crops. I had seen a few grainy photos, sent to me by my good friend and colleague Junchang Lü,* one of China's most famous dinosaur hunters. We both agreed it looked important. Maybe even one of those holy grail fossils—a new species, preserved in such an immaculate way that we can sense what it was like as a living, breathing creature tens of millions of years in the past. But we needed to see it ourselves to be sure.

When Junchang and I stepped off the train in Jinzhou, we were greeted by a band of local dignitaries, who took our bags and ushered us into two black SUVs. We were whizzed off to the city's museum, a surprisingly nondescript building on the outskirts of town. With the seriousness of a high-level political summit, we were led through the flickering neon lights of a long hallway, into a side room with a couple of desks and chairs. Balanced on a small table was a slab of rock so heavy that it seemed the legs were starting to buckle. One of the locals spoke in Chinese to Junchang, who then turned to me and gave a quick nod.

"Let's go," he said, in his curiously accented English, a combination of the Chinese cadence he grew up with and the Texas drawl that he picked up as a grad student in America.

The two of us came together and stepped toward the table. I could feel the eyes of everyone, an eerie silence hanging over the room as we approached the treasure.

Before me was one of the most beautiful fossils I had ever seen. It was a skeleton, about the size of a mule, its chocolate-brown bones standing out from the dull gray limestone surrounding it. A dinosaur for sure, its steak-knife teeth, pointy claws, and long tail leaving no doubt that it was a close cousin of *Jurassic Park*'s villainous *Velociraptor*.

* Junchang Lü tragically passed away in October 2018, leaving behind a legacy as one of the world's great dinosaur hunters.

But this was no ordinary dinosaur. Its bones were light and hollow, its legs long and skinny like a heron's, its slender skeleton the hallmark of an active, dynamic, fast-moving animal. And not only were there bones, but there were feathers covering the entire body. Bushy feathers that looked like hair on the head and neck, long branching feathers on the tail, and big quill pens on the arms, lined together and layered over each other to form wings.

This dinosaur looked just like a bird.

About a year later, Junchang and I described this skeleton as a new species, which we called *Zhenyuanlong suni*. It is one of about fifteen new dinosaurs that I've identified over the past decade, as I've carved out a career in paleontology that has taken me from my roots in the American Midwest to a faculty job in Scotland, with many stops all over the world to find and study dinosaurs.

Zhenyuanlong is unlike the dinosaurs I learned about in elementary school, before I became a scientist. I was taught that dinosaurs were big, scaly, stupid brutes so ill equipped for their environment that they just lumbered around, biding their time, waiting to go extinct. Evolutionary failures. Dead ends in the history of life. Primitive beasts that came and went, long before humans came on the scene, in a primeval world that was so different from today that it may as well have been an alien planet. Dinosaurs were curiosities to see in museums, or movie monsters that haunted our nightmares, or objects of childhood fascination, pretty much irrelevant to us today and unworthy of any serious study.

But these stereotypes are absurdly wrong. They've been dismantled over the past few decades, as a new generation has

Zhenyuanlong.

collected dinosaur fossils at an unprecedented rate. Somewhere around the world—from the deserts of Argentina to the frozen wastelands of Alaska—a new species of dinosaur is currently being found, on average, once a week. Let that sink in: a new dinosaur every . . . single . . . week. That's about fifty new species each year—*Zhenyuanlong* among them. And it's not only new discoveries but also novel ways of studying them—emerging technologies that help paleontologists understand the biology and evolution of dinosaurs in ways that our elders would have found unimaginable. CAT scanners are being used to study dinosaur brains and senses, computer models tell us how they moved, and high-power microscopes can even reveal what color some of them were. And so on.

It's been my great privilege to be part of this excitement—as one of many young paleontologists from across the globe, men

Junchang Lü and I studying the gorgeous fossil of *Zhenyuanlong*.

and women from many backgrounds who came of age in the era of *Jurassic Park*. There are a whole bunch of us twenty- and thirty-something researchers, working together and with our mentors from the preceding generation. With each new discovery we make, each new study, we learn a little more about dinosaurs and their evolutionary story.

That is the tale I am going to tell in this book—the epic account of where dinosaurs came from, how they rose to dominance, how some of them became colossal and others developed feathers and wings and turned into birds, and then how the rest of them disappeared, ultimately paving the way for the modern world, and for us. In doing so, I want to convey how we've pieced together this story using the fossil clues that we have, and give some sense of what it's like to be a paleontologist whose job it is to hunt for dinosaurs.

Most of all, though, I want to show that dinosaurs were not aliens, nor were they failures, and they're certainly not irrelevant. They were remarkably successful, thriving for over 150 million years and producing some of the most amazing animals that have ever lived—including birds, some ten thousand species of modern-day dinosaurs. Their home was our home—the same Earth, subject to the same whims of climate and environmental change that we have to deal with, or perhaps will deal with in the future. They evolved in concert with an ever changing world, one subject to monstrous volcanic eruptions and asteroid impacts, and one in which the continents were moving around, sea levels were constantly fluctuating, and temperatures were capriciously rising and falling. They became supremely well adapted to their environments, but in the end, most of them went extinct when they couldn't cope with a sudden crisis. No doubt there is a lesson there for us.

More than anything, the rise and fall of the dinosaurs is an incredible story, of a time when giant beasts and other fantastic creatures made the world their own. They walked on the very ground below us, their fossils now entombed in rock—the clues that tell this story. To me, it's one of the greatest narratives in the history of our planet.

STEVE BRUSATTE
Edinburgh, Scotland
MAY 18, 2017

1

The DAWN OF *the* DINOSAURS

Prorotodactylus

"BINGO," MY FRIEND GRZEGORZ NIEDŹWIEDZKI shouted, pointing at a knife-thin separation between a slim strip of mudstone and a thicker layer of coarser rock right above it. The quarry we were exploring, near the tiny Polish village of Zachełmie, was once a source of sought-after limestone but had long been abandoned. The surrounding landscape was littered with decaying smokestacks and other remnants of central Poland's industrial past. The maps deceitfully told us we were in the Holy Cross Mountains, a sad patch of hills once grand but now nearly flattened by hundreds of millions of years of erosion. The sky was gray, the mosquitoes were biting, heat was bouncing off the quarry floor, and the only other people we saw were a couple of wayward hikers who must have made a tragically wrong turn.

"This is the extinction," Grzegorz said, a big smile creasing the unshaven stubble of many days of fieldwork. "Many footprints of big reptiles and mammal cousins below, but then they disappear. And above, we see nothing for awhile, and then dinosaurs."

We may have been peering at some rocks in an overgrown quarry, but what we were really looking at was a revolution. Rocks record history; they tell stories of deep ancient pasts long before humans walked the Earth. And the narrative in front of us, written in stone, was a shocker. That switch in the rocks, detectable perhaps only to the overtrained eyes of a scientist, documents one of the most dramatic moments in Earth history. A brief instance when the world changed, a turning point that happened some 252 million years ago, before us, before woolly mammoths, before the dinosaurs, but one that still reverberates today. If things had unfolded a little differently back then, who knows what the modern world would be like? It's like wondering what might have happened if the archduke was never shot.

IF WE'D BEEN standing in this same spot 252 million years ago, during a slice of time geologists call the Permian Period, our surroundings would have been barely recognizable. No ruined factories or other signs of people. No birds in the sky or mice scurrying at our feet, no flowery shrubs to scratch us up or mosquitoes to feed on our cuts. All of those things would evolve later. We still would have been sweating, though, because it was hot and unbearably humid, probably more insufferable than Miami in the middle of the summer. Raging rivers would've been draining the Holy Cross Mountains, which were actually proper mountains back then, with sharp snowy peaks jutting tens of thousands of feet into the clouds. The rivers wound their way through vast forests of conifer trees—early relatives of today's pines and junipers—emptying into a big basin flanking the hills, dotted with lakes that swelled in the rainy season but dried out when the monsoons ended.

These lakes were the lifeblood of the local ecosystem, watering holes that provided an oasis from the harsh heat and wind. All sorts of animals flocked to them, but they weren't animals we would know. There were slimy salamanders bigger than dogs, loitering near the water's edge and occasionally snapping at a passing fish. Stocky beasts called pareiasaurs waddled around on all fours, their knobby skin, front-heavy build, and general brutish appearance making them seem like a mad reptilian offensive lineman. Fat little things called dicynodonts rummaged around in the muck like pigs, using their sharp tusks to pry up tasty roots. Lording over it all were the gorgonopsians, bear-size monsters who reigned at the top of the food chain, slicing into pareiasaur guts and dicynodont flesh with their saberlike canines. This cast of oddballs ruled the world right before the dinosaurs.

Then, deep inside, the Earth began to rumble. You wouldn't have been able to feel it on the surface, at least when it kicked off, right around 252 million years ago. It was happening fifty, maybe even a hundred, miles underground, in the mantle, the middle layer of the crust-mantle-core sandwich of Earth's structure. The mantle is solid rock that is so hot and under such intense pressure that, over long stretches of geological time, it can flow like extra-viscous Silly Putty. In fact, the mantle has currents just like a river. These currents are what drive the conveyor-belt system of plate tectonics, the forces that break the thin outer crust into plates that move relative to each other over time. We wouldn't have mountains or oceans or a habitable surface without the mantle currents. However, every once in a while, one of the currents goes rogue. Hot plumes of liquid rock break free and start snaking their way upward to the surface, eventually bursting out through volcanoes. These are called hot spots. They're rare, but Yellowstone is an example of an active one today. The constant supply of heat from the deep Earth is what powers Old Faithful and the other geysers.

This same thing was happening at the end of the Permian Period, but on a continent-wide scale. A massive hot spot began to form under Siberia. The streams of liquid rock rushed through the mantle into the crust and flooded out from volcanoes. These weren't ordinary volcanoes like the ones we're most used to, the cone-shaped mounds that sit dormant for decades and then occasionally explode with a bunch of ash and lava, like Mount Saint Helens or Pinatubo. They wouldn't have erupted with the vigor of those vinegar-and-baking-soda contraptions so many of us made as science fair experiments. No, these volcanoes were nothing more than big cracks in the ground, often miles long, that continuously belched out lava, year after year, decade after decade, century after century. The eruptions at the end of the Permian lasted for a few hundred thousand years, perhaps even a few million. There were a few bigger eruptive bursts and quieter periods of slower flow. All in all, they expelled enough lava to drown several million square miles of northern and central Asia. Even today, more than a quarter billion years later, the black basalt rocks that hardened out of this lava cover nearly a million square miles of Siberia, about the same land area as Western Europe.

Imagine a continent scorched with lava. It's the apocalyptic disaster of a bad B movie. Suffice it to say, all of the pareiasaurs, dicynodonts, and gorgonopsians living anywhere near the Siberian area code were finished. But it was worse than that. When volcanoes erupt, they don't expel only lava, but also heat, dust, and noxious gases. Unlike lava, these can affect the entire planet. At the end of the Permian, these were the real agents of doom, and they started a cascade of destruction that would last for millions of years and irrevocably change the world in the process.

Dust shot into the atmosphere, contaminating the high-altitude air currents and spreading around the world, blocking out the sun and preventing plants from photosynthesizing. The once lush conifer forests died out; then the pareiasaurs and dicynodonts had no plants to eat, and then the gorgonopsians had no meat. Food chains started to collapse. Some of the dust fell back through the atmosphere and combined with water droplets to form acid rain, which exacerbated the worsening situation on the ground. As more plants died, the landscape became barren and unstable, leading to massive erosion as mudslides wiped out entire tracts of rotting forest. This is why the fine mudstones in the Zachełmie quarry, a rock type indicative of calm and peaceful environments, suddenly gave way to the coarser boulder-strewn rocks so characteristic of fast-moving currents and corrosive storms. Wildfires raged across the scarred land, making it even more difficult for plants and animals to survive.

But those were just the short-term effects, the things that happened within the days, weeks, and months after a particularly large burst of lava spilled through the Siberian fissures. The longer-term effects were even more deadly. Stifling clouds of carbon dioxide were released with the lava. As we know all too well today, carbon dioxide is a potent greenhouse gas, which absorbs radiation in the atmosphere and beams it back down to the surface, warming up the Earth. The CO_2 spewed out by the Siberian eruptions didn't raise the thermostat by just a few degrees; it caused a runaway greenhouse effect that boiled the planet. But there were other consequences as well. Although a lot of the carbon dioxide went into the atmosphere, much of it also dissolved into the ocean. This causes a chain of chemical reactions that makes the ocean water more acidic, a bad thing, particularly for those sea creatures with easily dissolvable shells. It's why we don't bathe in vinegar. This chain reaction also draws much of the oxygen out of the oceans, another serious problem for anything living in or around water.

Descriptions of the doom and gloom could go on for pages, but the point is, the end of the Permian was a very bad time to be alive. It was the biggest episode of mass death in the history of our planet. Somewhere around 90 percent of all species disappeared. Paleontologists have a special term for an event like this, when huge numbers of plants and animals die out all around the world in a short time: a mass extinction. There have been five particularly severe mass extinctions over the past 500 million years. The one 66 million years ago at the end of the Cretaceous period, which wiped out the dinosaurs, is surely the most famous. We'll get to that one later. As horrible as the end-Cretaceous extinction was, it had nothing on the one at the end of the Permian. That moment of time 252 million years ago, chronicled in the swift change from mudstone to pebbly rock in the Polish quarry, was the closest that life ever came to being completely obliterated.

Then things got better. They always do. Life is resilient, and some species are always able to make it through even the worst catastrophes. The volcanoes erupted for a few million years, and then they stopped as the hot spot lost steam. No longer blighted by lava, dust, and carbon dioxide, ecosystems were gradually able to stabilize. Plants began to grow again, and they diversified. They provided new food for herbivores, which provided meat for carnivores. Food webs reestablished themselves. It took at least five million years for this recovery to unfold, and when it did, things were better but now very different. The previously dominant gorgonopsians, pareiasaurs, and their kin were never to stalk the lakesides of Poland or anywhere else while the plucky survivors had the whole Earth to themselves. A largely empty world, an uncolonized frontier. The Permian had transitioned into the next interval of geological time, the Triassic, and things would never be the same. Dinosaurs were about to make their entrance.

AS A YOUNG paleontologist, I yearned to understand exactly how the world changed as a result of the end-Permian extinction. What died and what survived, and why? How quickly did ecosystems recover? What new types of never-before-imagined creatures emerged from the post-apocalyptic blackness? What aspects of our modern world were first forged in the Permian lavas?

There's only one way to start answering these questions. You need to go out and find fossils. If a murder has been committed, a detective begins by studying the body and the crime scene, looking for fingerprints, hair, clothing fibers, or other clues that might tell the story of what unfolded, and lead to the culprit. For paleontologists, our clues are fossils. They are the currency of our field, the only records of how long-extinct organisms lived and evolved.

Fossils are any sign of ancient life, and they come in many forms. The most familiar are bones, teeth, and shells—the hard parts that form the skeleton of an animal. After being buried in sand or mud, these hard bits are gradually replaced by minerals and turned to rock, leaving a fossil. Sometimes soft things like leaves and bacteria can fossilize as well, often by making impressions in the rock. The same is sometimes true of the soft parts of animals, like skin, feathers, or even muscles and internal organs. But to end up with these as fossils, we need to be very lucky: the animal needs to be buried so quickly that these fragile tissues don't have time to decay or get eaten by predators.

Everything I describe above is what we call a body fossil, an actual part of a plant or animal that turns into stone. But there is another type: a trace fossil, which records the presence or behavior of an organism or preserves something that an organism produced. The best example is a footprint; others are burrows, bite marks, coprolites (fossilized dung), and eggs and nests. These can be particularly valuable, because they can tell us how extinct animals interacted with each other and their environment—how they moved, what they ate, where they lived, and how they reproduced.

The fossils that I'm particularly interested in belong to dinosaurs and the animals that came immediately before them. Dinosaurs lived during three periods of geological history: the Triassic, Jurassic, and Cretaceous (which collectively form the Mesozoic Era). The Permian Period—when that weird and wonderful cast of creatures was frolicking alongside the Polish lakes—came right before the Triassic. We often think of the dinosaurs as ancient, but in fact, they're relative newcomers in the history of life.

The Earth formed about 4.5 billion years ago, and the first microscopic bacteria evolved a few hundred million years later. For some 2 billion years, it was a bacterial world. There were no plants or animals, nothing that could easily be seen by the naked eye, had we been around. Then, some time around 1.8 billion years ago, these simple cells developed the ability to group together into larger, more complex organisms. A global ice age—which covered nearly the entire planet in glaciers, down to the tropics—came and went, and in its aftermath the first animals got their start. They were simple at first—soft sacs of goo like sponges and jellyfish, until they invented shells and skeletons. Around 540 million years ago, during the Cambrian Period, these skeletonized forms exploded in diversity, became extremely abundant, started eating one another, and began forming complex ecosystems in the oceans. Some of these animals formed a skeleton made of bones—these were the first vertebrates, and they looked like flimsy little minnows. But they, too, continued to diversify and eventually some of them turned their fins into arms, grew fingers and toes, and emerged onto the land, about 390 million years ago. These were the first tetrapods, and their descendants include all vertebrates that live on land today: frogs and salamanders, crocodiles and snakes, and then later, dinosaurs and us.

We know this story because of fossils—thousands of skeletons and teeth and footprints and eggs found all over the world by generations of paleontologists. We're obsessed with finding fossils and notorious for going to great (and sometimes stupid) lengths to discover new ones. It could be a limestone pit in Poland or maybe a bluff behind a Walmart, a dump pile of boulders at a construction site, or the rocky walls of a ripe landfill. If there are fossils to be found, then at least some swashbuckling (or stupid) paleontologist will brave whatever heat, cold, rain, snow, humidity, dust, wind, bug, stench, or war zone stands in the way.

That's why I started going to Poland. I first visited in the summer of 2008, a twenty-four-year-old in between finishing my master's and starting my PhD; I went to study some intriguing new reptile fossils that had been found a few years earlier in Silesia, the sliver of southwestern Poland that for years was fought over by Poles, Germans, and Czechs. The fossils were kept in a museum in Warsaw, treasures of the Polish state. I remember the buzz as I approached the capital's central station on a delayed train from Berlin, night shadows covering the hideous Stalin-era architecture of a city rebuilt from ruins after the war.

As I stepped off the train, I scanned the crowd. Somebody was supposed to be there holding a sign with my name. I arranged my visit through a series of formal e-mails with a very senior Polish professor, who badgered one of his graduate students into meeting me at the station and guiding me to the small guestroom where I would stay at the Polish Institute of Paleobiology, just a few stories above where the fossils were kept. I had no idea whom I was looking for, and because the train had been more than an hour late, I figured the student had escaped back to the lab, leaving me on my own to navigate a foreign city in the twilight, with the few words of Polish on the glossary page of my guidebook.

Just as I was starting to panic, I saw a sheet of white paper flapping in the wind, my name hastily scrawled across it. The man holding it was young, with a close-cropped military hairstyle, his hairline just starting to recede like mine. His eyes were dark, and he was squinting. A thin veneer of stubble covered his face, and he seemed to be a little darker than most of the Poles I knew. Tanned, almost. There was something vaguely sinister about him, but that changed in an instant when he recognized me coming toward him. He broke into a huge smile, grabbed my bag, and gripped my hand firmly. "Welcome to Poland. My name is Grzegorz. How about some dinner?"

We were both tired, I from the long train journey, Grzegorz from working the whole day describing a new batch of fossil bones that he and his crew of undergraduate assistants had just found in southeastern Poland a few weeks before, hence the field tan he was sporting. But we ended up knocking back several beers and talking for hours about fossils. This guy had the same raw enthusiasm for dinosaurs that I had, and he was full of iconoclastic ideas about what happened after the end-Permian extinction.

Grzegorz and I became fast friends. For the rest of that week, we studied Polish fossils together, and then during the following four summers, I came back to Poland to do fieldwork with Grzegorz, often joined by the third musketeer in our band, the young British paleontologist Richard Butler. During that time we found a lot of fossils and came up with some new ideas about how dinosaurs got their evolutionary start in those heady days after the end-Permian extinction. Over the course of those years, I saw Grzegorz transition from an eager, but still somewhat meek, graduate student into one of Poland's leading paleontologists. A few years before turning thirty, he discovered, in a different corner of the Zachełmie quarry, a trackway left by one of those first fishy creatures to walk out of the water and onto land, some 390 million years ago. His discovery was published on the cover of *Nature*, one of the world's leading scientific journals. He was invited to a special audience with Poland's prime minister and gave a TED talk. His steely face—not his fossil discoveries, *him*—graced the cover of the Polish version of *National Geographic*.

He had become something of a scientific celebrity, but more than anything else Grzegorz enjoyed heading out into nature and looking for fossils. He called himself a "field animal," explaining that he loved camping and hacking through brush much more than the genteel ways of Warsaw. He couldn't help it. He grew up around Kielce, the main city of the Holy Cross Mountains region, and started collecting fossils as a child. He developed a particular talent for finding a type that many paleontologists ignore: trace fossils. Footprints, hand impressions, tail drags: the marks dinosaurs and other animals left when they moved across mud or sand, going about their daily business of hunting, hiding, mating, socializing, feeding, and loitering. He was absolutely enamored of tracks. An animal has only one skeleton, but it can leave millions of footprints, he would often remind me. Like an intelligence operative, he knew all the best places to find them. This was his backyard, after all. It was quite the backyard to grow up in, too, because it turned out that those animal-infested seasonal lakes that covered the area during the Permian and Triassic were perfect environments for preserving tracks.

For four summers we indulged Grzegorz's love of tracks. Richard and I tagged along as he led us to many of his secret sites, which were mostly abandoned quarries, bits of rock poking out of streams, and rubbish piles along the ditches of the many new roads that were being built in the area, where workmen would dump the slabs of stone they cut through when laying asphalt. We found a lot. Or rather, Grzegorz did. Both Richard and I developed an eye for the often small hand- and footprints left by lizards, amphibians, and early dinosaur and crocodile relatives, but we could never compete with the master.

The thousands of tracks that Grzegorz found over his two decades of collecting, plus the pittance of new ones that Richard and I stumbled upon, ended up telling quite a story. There were many types of tracks, belonging to a whole slew of different creatures. And they didn't come from just one moment in time, but from a sequence of tens of millions of years, beginning in the Permian, continuing across the great extinction into the Triassic, and even reaching the next stage of geological time, the Jurassic Period, which began about 200 million years ago. When the seasonal lakes dried up, they left vast mud flats that animals walked across, leaving their marks. The rivers would continuously bring in new sediment to cover up the mud flats, burying them and turning them to stone. The cycle repeated year after year after year, so there is now layer upon layer upon layer of tracks in the Holy Cross Mountains. For paleontologists this is a bonanza: an opportunity to see how animals and ecosystems were changing over time, particularly after the cataclysmic end-Permian extinction.

Identifying what animals made which particular track is relatively straightforward. You compare the shape of the track to the shape of hands and feet. How many fingers or toes are there? Which ones are longest? Which way do they face? Do only the fingers and toes make an impression, or does the palm of the hand and arch of the foot also leave a mark? Are the left and right tracks really close together, as the trackmaker was walking with its limbs right under its body, or are they far apart, made by a creature with limbs sprawled out to the side? By following this checklist, you can usually figure out which general group of animals left the tracks in question. Pinpointing an exact species is almost impossible, but distinguishing the tracks of reptiles from amphibians, or dinosaurs from crocodiles, is easy enough.

The Permian tracks from the Holy Cross Mountains are a diverse lot, and most were made by amphibians, small reptiles, and early synapsids, progenitors of mammals that are often annoyingly, and incorrectly, described as mammal-like reptiles (although they are not actually reptiles) in kids' books and museum exhibits. Gorgonopsians and dicynodonts are two types of these primitive synapsids. By all accounts these latest Permian ecosystems were strong—there were many varieties of animals, some small and others more than ten feet long and weighing over a ton, living together, thriving in the arid climate along the seasonal lakes. There are, however, no signs of dinosaur or crocodile tracks in the Permian layers, or even any tracks that look like precursors to these animals.

Everything changes at the Permian-Triassic boundary. Following the tracks across the extinction is like reading an arcane book in which a chapter of English follows one written in Sanskrit. The latest Permian and earliest Triassic seem to be two different worlds, which is remarkable because the tracks were all left in the identical place, in the same exact environment and climate. Southern Poland didn't stop being a humid lakeland fed by raging mountain streams as the Permian ticked over into the Triassic. No, it was the animals themselves that changed.

I get the creeps when looking at the earliest Triassic tracks. I can sense the long-distant specter of death. There are hardly any tracks at all, just a few small prints here and there, but a lot of burrows jutting deep into the rock. It seems the surface world was annihilated and whatever creatures inhabited this haunted landscape were hiding underground. Almost all of the tracks belong to small lizards and mammal relatives, probably not much larger than a groundhog. Many of the diverse tracks of the Permian are gone, particularly those made by the larger proto-mammal synapsids, and they never reappear.

Things gradually start to improve as you follow the tracks up through time. More track types appear, some of the prints get larger, and burrows become rarer. The world was clearly recovering from the shock of end-Permian volcanoes. Then, about 250 million years ago, just a couple of million years after the extinction, a new type of track starts showing up. They're small, just a few centimeters long, about the size of a cat's paw. They are arranged in narrow trackways, the five-fingered handprints positioned in front of the slightly larger footprints, which have three long central toes flanked by a tiny toe on each side. The best place to find them is near a tiny Polish village called Stryczowice, where you can park your car at a bridge, scramble your way through thorns and bramble, and poke around the banks of a narrow stream littered with track-covered rock slabs. Grzegorz discovered the site when he was young and proudly took me there once, on a miserable July day of obscene humidity, bugs, rain, and thunder. After a few minutes of hacking through the weeds, we were soaked, my field notebook warping as ink started to run off the pages.

The tracks found here go by the scientific name of *Prorotodactylus*. Grzegorz wasn't quite sure what to make of them. They were certainly different from the other tracks found alongside them, and all of the tracks from the Permian. But what kind of animal made them? Grzegorz had a hunch they could have something to do with dinosaurs, because an elderly paleontologist named Hartmut Haubold had reported similar tracks from Germany in the 1960s and had argued that they were made by early dinosaurs or close cousins. But Grzegorz wasn't sold on the idea. He had spent most of his young career studying tracks and hadn't spent much time with actual dinosaur skeletons, so it was difficult for him to match the prints to a trackmaker. That's where I came in. For my master's degree, I constructed a family tree of Triassic reptiles, a genealogy showing how the first dinosaurs were related to the other animals of the time. I spent months in museum collections studying fossil bones, so I knew the anatomy of the first dinosaurs quite well. As did Richard, who wrote a PhD thesis on early dinosaur evolution. The three of us put our heads together to figure out what culprit was responsible for the *Prorotodactylus* tracks, and we did indeed conclude that it was a very dinosaurlike animal. We announced our interpretation in a scientific paper we published in 2010.

The clues, of course, are in the details of the tracks. When I look at the *Prorotodactylus* trackways, the first thing that jumps out at me is that they are very narrow. There is only a little bit of space between the left and right tracks in the sequence, just a few centimeters. There's only one way for an animal to make tracks like this: by walking upright, with the arms and legs right underneath the body. We walk upright, so when we leave footprints on the beach, the left and right ones are very close together. Same with a horse—take a look at the pattern of horseshoe impressions left by a galloping horse next time you're on a farm (or wagering a few bucks at the track), and you'll see what I mean. But this style of walking is actually quite rare in the animal kingdom. Salamanders, frogs, and lizards move in a different way. Their arms and legs stick out sideways from the body. They sprawl. That means their trackways are much wider, with big separation between the left and right tracks made by their spread-eagle limbs.

The Permian world was dominated by sprawlers. After the extinction, however, one new group of reptiles evolved from these sprawlers but developed an upright posture—the archosaurs. This was a landmark evolutionary event. Sprawling is all well and good for cold-blooded critters that don't need to move very fast. Tucking your limbs under your body, however, opens up a new world of possibilities. You can run faster, cover greater distances, track down prey with greater ease, and do it all more efficiently, wasting less energy as your columnar limbs move

Grzegorz Niedźwiedzki examines a life-size model of the *Prorotodactylus* trackmaker: a proto-dinosaur very similar to the ancestor that gave rise to dinosaurs. *Courtesy of Grzegorz Niedźwiedzki.*

A handprint overlapping a footprint of *Prorotodactylus*, from Poland. For scale, the handprint is about 1 inch long.

back and forth in an orderly fashion rather than twisting around like those of a sprawler.

We may never know exactly why some of these sprawlers started walking upright, but it probably was a consequence of the end-Permian extinction. It's easy to imagine how this new getup gave archosaurs an advantage in the postextinction chaos, when ecosystems were struggling to recover from the volcanic haze, temperatures were unbearably hot, and empty niches abounded, waiting to be filled by whatever mavericks could evolve ways to endure the hellscape. Walking upright, it seems, was one of the ways in which animals recovered—and indeed, improved—after the planet was shocked by the volcanic eruptions.

Not only did the new upright-walking archosaurs endure, but they thrived. From their humble origins in the traumatic world of the Early Triassic, they later diversified into a staggering variety of species. Very early, they split into two major lineages, which would grapple with each other in an evolutionary arms race over the remainder of the Triassic. Remarkably, both of these lineages survive today. The first, the pseudosuchians, later gave rise to crocodiles. As shorthand, they are usually referred to as the crocodile-line archosaurs. The second, the avemetatarsalians, developed into pterosaurs (the flying reptiles often called pterodactyls), dinosaurs, and by extension the birds that, as we shall see, descended from the dinosaurs. This group is called the bird-line archosaurs. The *Prorotodactylus* tracks from Stryczowice are some of the first signs of archosaurs in the fossil record, traces of the great-great-great-grandmother of this whole menagerie.

Exactly what kind of archosaur was *Prorotodactylus*? Some peculiarities in the footprints hold important clues. Only the toes make an impression, not the metatarsal bones that form the arch of the foot. The three central toes are bunched very close together, the two other toes are reduced to nubbins, and the back end of the print is straight and razor-sharp. These may seem like anatomical minutiae, and in many ways they are. But as a doctor is able to diagnose a disease from its symptoms, I can recognize these features as hallmarks of dinosaurs and their very closest cousins. They link to unique features of the dinosaur foot skeleton: the digitigrade setup, in which only the toes make contact with the ground when walking, the very narrow foot in which the metatarsals and toes are bunched together, the pathetically atrophied outer toes, the hinge-like joint between the toes and the metatarsals, which reflects the characteristic ankle of dinosaurs and birds, which can move only in a back-and-forth direction, without even the slightest possibility of twisting.

The *Prorotodactylus* tracks were made by a bird-line archosaur very closely related to the dinosaurs. In scientific parlance, this makes *Prorotodactylus* a dinosauromorph, a member of that group that includes dinosaurs and the handful of their very closest cousins, those few branches just below the bloom of dinosaurs on the family tree of life. After the evolution of the upright-walking archosaurs from the sprawlers, the origin of dinosauromorphs was the next big evolutionary event. Not only did these dinosauromorphs stand proudly on their erect limbs, but also they had long tails, big leg muscles, and hips with extra bones connecting the legs to the trunk, all of which allowed them to move even faster and more efficiently than other upright-walking archosaurs.

As one of the first dinosauromorphs, *Prorotodactylus* is something of a dinosaur version of Lucy, the famous fossil from Africa that belongs to a very humanlike creature but is not quite a true human, a member of our species, *Homo sapiens*. In the same way that Lucy looks like us, *Prorotodactylus* would have appeared and behaved very much like a dinosaur, but it's simply not considered a true dinosaur by convention. That's because scientists decided long ago that a dinosaur should be defined as any members belonging to that group including the plant-eating *Iguanodon* and the meat-eating *Megalosaurus* (two of the first dinosaurs found by scientists in the 1820s) and all descendants of their common ancestor. Because *Prorotodactylus* did not evolve from this common ancestor, but slightly before it, it is not a true dinosaur by definition. But that's just semantics.

In *Prorotodactylus* we're looking at traces left behind by the type of animal that evolved into dinosaurs. It was about the size of a house cat and would have been lucky to tip the scales at ten pounds. It walked on all fours, leaving handprints and footprints. Its limbs must have been quite long, judging from the big gaps between successive prints of the same hands and feet. The legs must have been particularly long and skinny, because the footprints often are positioned in front of the handprints, a sign that its feet were overstepping its hands. The hands were small and would have been good at grabbing things, whereas the long, compressed feet were perfect for running. The *Prorotodactylus* animal would have been gangly looking, with the speed of a cheetah but the awkward proportions of a sloth, perhaps not the type of animal you would expect the great *Tyrannosaurus* and *Brontosaurus* to ultimately evolve from. And it wasn't very common either: less than 5 percent of all the tracks found at Stryczowice belong to *Prorotodactylus*, an indication that these proto-dinosaurs were not especially abundant or successful when they first arose. Instead, they were far outnumbered by small reptiles, amphibians, and even other types of primitive archosaurs.

These rare, weird, not-quite-true-dinosaur dinosauromorphs continued to evolve as the world healed in the Early and Middle Triassic. The Polish track sites, stacked orderly in time sequence like the pages of a novel, document it all. Sites like Wióry, Pałęgi, and Baranów yield an equally unfamiliar array of dinosauromorph tracks—*Rotodactylus*, *Sphingopus*, *Parachirotherium*, *Atreipus*—which diversify over time. More and more track types show up; they get larger; they develop a greater diversity of shape, some even losing their outer toes entirely so that the center toes are all that remain. Some of the trackways stop showing impressions of the hand—these dinosauromorphs were walking on only their hind legs. By about 246 million years ago, dinosauromorphs the size of wolves were racing around on two legs, grabbing prey with their clawed hands, acting a whole lot like a pint-size version of a *T. rex*. They weren't living only in Poland; their footprints are also found in France and Germany and the southwestern United States, and their bones start showing up in eastern Africa and later Argentina and Brazil. Most of them ate meat, but some of them turned vegetarian. They moved quickly, grew fast, had high metabolisms, and were active, dynamic animals compared to the lethargic amphibians and reptiles they were cohabitating with.

At some point, one of these primitive dinosauromorphs evolved into true dinosaurs. It was a radical change in name only. The boundary between nondinosaurs and dinosaurs is fuzzy, even artificial, a by-product of scientific convention. The same way that nothing really changes as you cross the border from Illinois into Indiana, there was no profound evolutionary leap as one of these dog-size dinosauromorphs changed into another dog-size dinosauromorph that was just over that dividing line on the family tree that denotes dinosaurs. This transition involved the development of only a few new features of the skeleton: a long scar on the upper arm that anchored muscles to move the arms in and out, some tablike flanges on the neck vertebrae that supported stronger muscles and ligaments, and an open-window-like joint where the thighbone meets the pelvis. These were minor changes, and to be honest, we don't really know what was driving them, but we know that the dinosauromorph-dinosaur transition wasn't a major evolutionary jump. A far bigger evolutionary event was the origin of the swift-running, strong-legged, fast-growing dinosauromorphs themselves.

The first true dinosaurs arose some time between 240 and 230 million years ago. The uncertainty reflects two problems that continue to cause me headaches but are ripe to be solved by the next generation of paleontologists. First, the earliest dinosaurs are so similar to their dinosauromorph cousins that it is hard to tell their skeletons apart, never mind their footprints. For instance, the puzzling *Nyasasaurus*, known from part of an arm and a few vertebrae from approximately 240-million-year-old rocks in Tanzania, may be the world's oldest dinosaur. Or it may be just another dinosauromorph on the wrong side of the genealogical divide. The same is true of some of the Polish footprints, particularly the larger ones made by animals walking on their hind legs. Maybe some of these were made by real, true, honest-to-goodness dinosaurs. We just don't have a good way of telling apart the tracks of the earliest dinosaurs and their closest nondinosaur relatives, because their foot skeletons are so similar. But maybe it doesn't matter too much, as the origin of true dinosaurs was much less important than the origin of dinosauromorphs.

The other, much more glaring issue is that many of the fossil-bearing rocks of the Triassic are very poorly dated, particularly those from the early to middle parts of the period. The best way to figure out the age of rocks is to use a process called radiometric dating, which compares the percentages of two different types of elements in the rock—say, potassium and argon. It works like this. When a rock cools from a liquid into a solid, minerals form. These minerals are made up of certain elements, in our case including potassium. One isotope (atomic form) of potassium (potassium-40) is not stable, but slowly undergoes a process called radioactive decay, in which it changes into argon-40 and expels a small amount of radiation, causing the beeps you'd hear on a Geiger counter. Beginning the moment a rock solidifies, its unstable potassium starts changing into argon. As this process continues, the accumulating argon gas becomes trapped inside the rock where it can be measured. We know from lab experiments the rate at which potassium-40 changes into argon-40. Knowing this rate, we can take a rock, measure the percentages of the two isotopes, and calculate how old the rock is.

Radiometric dating revolutionized the field of geology in the middle of the twentieth century; it was pioneered by a Brit named Arthur Holmes, who once occupied an office a few doors down from mine at the University of Edinburgh. Today's labs, like the ones run by my colleagues at New Mexico Tech and the Scottish Universities Environmental Research Centre near Glasgow, are high-tech, ultramodern facilities where scientists in white lab coats use multi million-dollar machines bigger than my old Manhattan apartment to date microscopic rock crystals. The techniques are so refined that rocks hundreds of millions of years old can be precisely dated to a small window of time, within a few tens or hundreds of thousands of years. These methods are so fine-tuned that independent labs routinely calculate the same dates for samples of the same rocks analyzed blindly. Good scientists check their work this way, to make sure their methodology is sound, and test after test has shown that radiometric dating is accurate.

But there is one major caveat: radiometric dating works only on rocks that cool from a liquid melt, like basalts or granites that solidify from lava. The rocks that contain dinosaur fossils, like mudstone and sandstone, were not formed this way, but rather from wind and water currents that dumped sediment. Dating these types of rocks is much more difficult. Sometimes a paleontologist is lucky and finds a dinosaur bone sandwiched between two layers of datable volcanic rocks that provide a time envelope for when that dinosaur must have lived. There are other methods that can date individual crystals found in sandstones and mudstones, but these are expensive and time-consuming. This means that it's often difficult to date dinosaurs accurately. Some parts of the dinosaur fossil record have been well dated—when there are enough interspersed volcanic rocks to give a timeline or the individual-crystal technique has been successful—but not the Triassic. There are just a handful of well-dated fossils, so we are not entirely confident of what order certain dinosauromorphs appeared in (especially when trying to compare the ages of species found in distant parts of the world) or when true dinosaurs emerged out of the dinosauromorph stock.

ALL UNCERTAINTIES ASIDE, we do know that by 230 million years ago, true dinosaurs had entered the picture. The fossils of several species with unquestionable signature features of dinosaurs are found in well-dated rocks of that age. They're found in a place far from where the earliest dinosauromorphs were cavorting in Poland—the mountainous canyons of Argentina.

Ischigualasto Provincial Park, in the northeastern part of Argentina's San Juan Province, is the type of place that just looks as though it should be bursting with dinosaurs. It's also called Valle de la Luna—the Valley of the Moon—and you could easily imagine its being on some other planet, full of wind-sculpted hoodoos, narrow gullies, rust-covered cliffs, and dusty badlands. To the northwest are the towering peaks of the Andes, and far to the south are the dry plains that cover most of the country, where cows graze on the grass that makes Argentine beef so delicious. For centuries Ischigualasto has been an important crossing for livestock making their way from Chile to Argentina, and today many of the few people who live in the area are ranchers.

This stunning landscape also happens to be the best place in the world for finding the oldest dinosaurs. That's because the red, brown, and green rocks that have been carved and eroded into such magical shapes were formed in the Triassic, in an environment both full of life and perfect for preserving fossils. In many ways, this landscape was similar to the Polish lakelands that preserved the tracks of *Prorotodactylus* and other dinosauromorphs. The climate was hot and humid, although perhaps a little more arid and not pounded by such strong seasonal monsoons. Rivers snaked their way into a deep basin, occasionally bursting their banks during rare storms. Over a period of 6 million years, the rivers built up repeating sequences of sandstone, formed in the river channels, and mudstone, formed from the finer particles that escaped the river and settled out on the surrounding floodplains. Many dinosaurs frolicked on these plains, along with a wealth of other animals—big amphibians, piglike dicynodonts whose ancestors managed to make it through the end-Permian extinction, beaked plant-eating reptiles called rhynchosaurs (primitive cousins of the archosaurs) and furry little cynodonts that looked like a cross between a rat and an iguana. Floods would occasionally interrupt this paradise, killing the dinosaurs and their friends and burying their bones.

The area is so heavily eroded today, and so little disturbed by buildings and roads and other human nuisances that cover up fossils, that the dinosaurs are relatively easy to find, at least compared to so many other parts of the world where we hike around for days just praying to find anything, even just a tooth. The very first discoveries here were made by cowpokes or other locals, and it wasn't until the 1940s that scientists began to collect, study, and describe fossils from Ischigualasto, then still another few decades until intensive expeditions were launched.

The first major collecting trips were led by one of the giants of twentieth-century paleontology, the Harvard professor Alfred Sherwood Romer, the man who wrote *the* textbook that I still use to teach my graduate students in Edinburgh. During his first trip, in 1958, Romer was already sixty-four years old and regarded as a living legend, yet there he was driving a rickety car through the badlands because he had a hunch that Ischigualasto would be the next big frontier. On that trip he found part of a skull and skeleton of a "moderately large" animal, as he so modestly put it in his field notebook. He brushed away as much rock as he could, coated the bones in newspaper, applied a coat of plaster that would harden and protect the bones, and chiseled them out of the ground. He sent the bones back to Buenos Aires, where they would be loaded on a ship to the United States, so he could carefully clean and study them in his lab. But the fossils took a detour. They were impounded for two years at the port in Buenos Aires before customs officials finally gave the go-ahead. By the time the fossils arrived at Harvard, Romer had occupied himself with other things, and it was only years later that other paleontologists recognized that the master had found the very first good dinosaur from Ischigualasto.

Some Argentines weren't so happy that a Norteamericano had come down to their neighborhood to collect fossils, which were being removed from Argentina and studied in the United States. That spurred a pair of up-and-coming homegrown scientists, Osvaldo Reig and José Bonaparte, to organize their own expeditions. They assembled a team and set out for Ischigualasto in 1959, and then again three times during the early 1960s. It was during the 1961 field season that Reig and Bonaparte's crew met a local rancher and artist named Victorino Herrera, who knew the hills and crevasses of Ischigualasto the way an Inuit knows snow. He recalled seeing some bones crumbling out of the sandstone and led the young scientists to the spot.

Herrera had found bones all right, lots of them, and clearly they were part of the back end of a dinosaur skeleton. After a few years of study, Reig described the fossils as a new species of dinosaur that he called *Herrerasaurus* in the rancher's honor, a mule-size creature that could sprint on its hind legs. Later detective work showed that Romer's impounded fossils belonged to the same animal, and future discoveries revealed that *Herrera-*

saurus was a fierce predator with an arsenal of sharp teeth and claws, a primitive version of *T. rex* or *Velociraptor. Herrerasaurus* was one of the very first theropod dinosaurs—a founding member of that dynasty of smart, agile predators that would later ascend to the top of the food chain and ultimately evolve into birds.

You might think this discovery would have encouraged paleontologists from throughout Argentina to flock to Ischigualasto in some kind of mad dinosaur rush. But it didn't happen. After Reig and Bonaparte's expeditions ended, things got quiet. The late 1960s and 1970s were not a prime time for dinosaur research. There was little funding and, believe it or not, little public interest. Things picked up again in the late 1980s, when a thirty-something paleontologist from Chicago named Paul Sereno put together a joint Argentine-American team of other ambitious young guns, mostly graduate students and junior professors. They set out in the footsteps of Romer, Reig, and Bonaparte, the latter meeting with the group for a few days to guide them to some of his favorite sites. The trip was a rousing success: Sereno found another skeleton of *Herrerasaurus* and many other dinosaurs, proving that Ischigualasto still had plenty of fossils to give up.

Three years later, Sereno was at it again, bringing much of the same crew back to Ischigualasto to explore new territory. One of his assistants was a wisecracking student named Ricardo Martínez. While out prospecting one day, Martínez picked up a fist-size hunk of rock covered in a gnarly frosting of iron minerals. *Just another piece of junk*, he thought, but as he reached back to toss it aside, Martínez noticed something pointy and shiny sticking out of the cobble. They were teeth. Glancing back at the ground, dumbfounded, he realized that he had plucked the head off the nearly complete skeleton of a dinosaur, a long-legged, lightly built speed demon about the size of a golden retriever. They named it *Eoraptor.* Those teeth poking out from the skull turned out to be highly unusual: the ones in the back of the jaw were sharp and serrated like a steak knife, surely to slice through flesh, but the ones at the tip of the snout were leaf-shaped with coarse projections called denticles, the same type of tooth that some long-necked, potbellied sauropod dinosaurs would later use to grind plants. This hinted that *Eoraptor* was an omnivore and possibly a very early member of the sauropod lineage, a primitive cousin of *Brontosaurus* and *Diplodocus.*

I met Ricardo Martínez many years later, around the time that I first laid eyes on the gorgeous skeleton of *Eoraptor.* I was an undergraduate student at the University of Chicago, training in Paul Sereno's lab, when Ricardo came to work on a clandestine project, later announced as yet another new dinosaur from Ischigualasto, the terrier-size primitive theropod *Eodromaeus.* I took a liking to Ricardo right away. Paul was running an hour late, stuck in traffic on Lake Shore Drive, and Ricardo was literally twiddling his thumbs, hunched in the corner of the lab office. It was an incongruously disengaged posture from a man who very quickly revealed himself to be the very type of hot-blooded, fast-talking, fossil-loving typhoon that I longed to be. He kind of looked like the Dude from *The Big Lebowski*: wild tangled hair, beard thick around the mouth, interesting fashion sense. He regaled me with stories of working in the wilds of Argentina, recounting with theatrical hand gestures how his hungry crew would sometimes hunt down stray cattle on their ATVs, delivering killing blows with the business end of their geological rock hammers. He could tell I was developing a romantic attraction to Argentina and told me to look him up if I ever came to visit.

Five years later, I took him up on the offer when I attended the hardest-rocking scientific conference I've ever had the pleasure of speaking at. Usually conferences are fairly stale affairs, held in Marriotts and Hyatts in cities like Dallas and Raleigh, where scientists gather to listen to each other speak in cavernous banquet halls that usually host weddings, drinking overpriced hotel beer while catching up on field stories. The conference that Ricardo and his colleagues hosted in the city of San Juan was anything but. The dinner on the last evening was legendary, like one of those hedonistic house parties in a rap video. A local politician adorned with a sash opened the proceedings, managing to make an outrageous quip about some of the foreigners in attendance. The main course was a phonebook-size slab of grass-fed beef, washed down with copious amounts of red wine. After dinner was dancing, for hours, fueled by an open bar with hundreds of bottles of vodka, whiskey, brandy, and a local firewater whose name I can't remember. At about three A.M., there was a break in the proceedings while a make-your-own taco bar was assembled outside, a tasty change from the humidity of the dance floor. We staggered back to our hotels as dawn broke. Ricardo was right. I would love Argentina.

Before the debauchery of that evening, I spent several days in the collections of Ricardo's museum, the Instituto y Museo de Ciencias Naturales in the lovely city of San Juan. Most of the riches of Ischigualasto are kept here, *Herrerasaurus*, *Eoraptor*, and *Eodromaeus* among them, but also many other dinosaurs. There's *Sanjuansaurus*, a close cousin of *Herrerasaurus* that was also a fierce predator. In another drawer is *Panphagia*, similar to *Eoraptor* in being a primitive miniature cousin of the later colossal sauropods, and *Chromogisaurus*, a larger *Brontosaurus* relative that grew up to a couple of meters long and was something of a middle-of-the-food-chain plant-eater. There are also the scrappy fossils of a dinosaur called *Pisanosaurus*, a dog-size animal that shares some features of the teeth and jaws with the ornithischian dinosaurs—the group that would later diversify into a vast range of plant-eating species, from the horned *Triceratops* to the duck-billed hadrosaurs. And they're still finding new dinosaurs in Ischigualasto, so who knows what new characters will be added if you are lucky enough to visit.

As I was pulling open the specimen cabinet doors, carefully removing the fossils to measure and photograph them, I felt like something of an historian, one of those scholars who spends dark hours in the archives, scrutinizing ancient manuscripts. The analogy is deliberate, because the Ischigualasto fossils are indeed historical artifacts, primary-source objects that help us tell the story of deep prehistoric pasts, millions of years before monks started writing on parchment. The bones that Romer, Reig, and Bonaparte, and then later Paul, Ricardo, and their many colleagues, have pried from the lunar landscape of Ischigualasto are the very first records of true dinosaurs, living, evolving, and beginning their long march to dominance.

These first dinosaurs weren't quite dominant yet, overshadowed by the larger and more diverse amphibians, mammal cousins, and crocodile relatives that they lived alongside on those dry, occasionally flooded plains of the Triassic. Even *Herrerasaurus* probably wasn't at the top of the food chain, ceding that title to the murderous twenty-five-foot-long crocodile-line

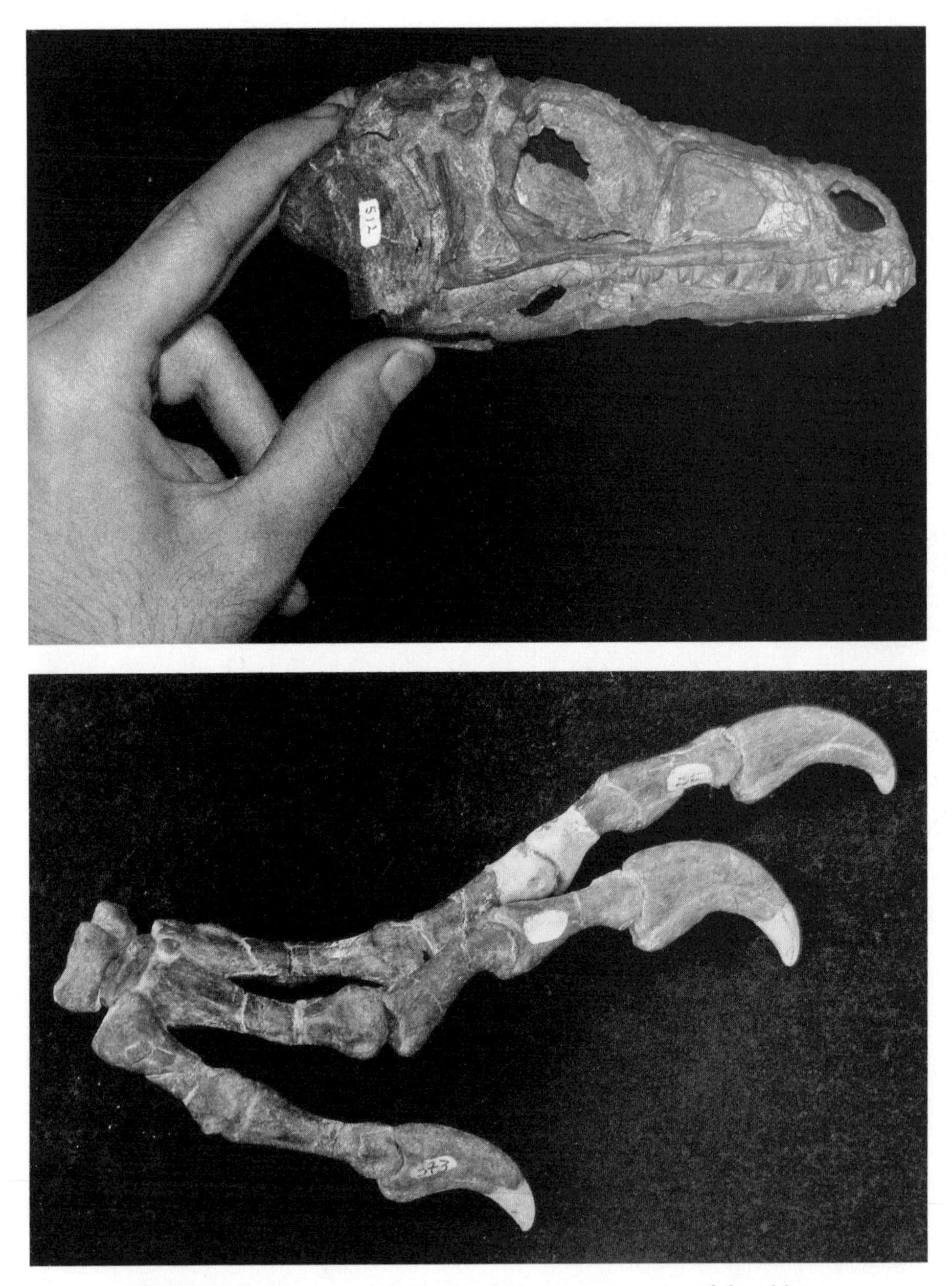

The skull of *Eoraptor* and the hand of *Herrerasaurus*, two of the oldest dinosaurs.

archosaur *Saurosuchus*. But the dinosaurs had arrived on the scene. The three major groups—the meat-eating theropods, long-necked sauropods, and herbivorous ornithischians—had already diverged from each other on the family tree, siblings setting out to form their own broods.

The dinosaurs were on the march.

2

DINOSAURS RISE UP

Coelophysis

IMAGINE A WORLD WITH NO BORDERS. I'm not channeling John Lennon. What I mean is, envision a version of Earth where all of the land is connected together—no patchwork of continents separated by oceans and seas, just a single expanse of dry ground stretching from pole to pole. Given enough time and a good pair of shoes, you could walk from the Arctic Circle across the equator to the South Pole. If you ventured too far inland, you would find yourself many thousands of miles—tens of thousands, even—from the closest beach. But if you fancied a swim, you could take a dip in the vast ocean surrounding the big slab of land you called home and, theoretically at least, paddle from one coast all the way around the planet to the other coast without having to dry off.

It may sound fanciful, but this is the world the dinosaurs grew up in.

When the very first dinosaurs, like *Herrerasaurus* and *Eoraptor*, evolved from their cat-size dinosauromorph ancestors some 240 to 230 million years ago, there were no individual continents—no Australia or Asia or North America. There was no Atlantic Ocean separating the Americas from Europe and Africa, no Pacific Ocean on the flip side of the globe. Instead, there was just one huge solid unbroken mass of land—what geologists refer to as a supercontinent. It was surrounded by a single global ocean. Geography class would have been easy in those days: the supercontinent we call Pangea, and the ocean we call Panthalassa.

The dinosaurs were born into what we would see as a totally alien world. What was it like to live in such a place?

First, let's think about the physical geography. The supercontinent spanned an entire hemisphere of the Triassic Earth from

North Pole to South. It looked something like a gigantic letter C, with a big indentation in the middle where an arm of Panthalassa cut into the land. Towering mountain ranges snaked across the landscape at odd angles, marking the sutures where smaller blocks of crust had once collided to build the giant continent, the pieces of a jigsaw puzzle. This puzzle wasn't put together very easily or very quickly. For hundreds of millions of years, heat deep inside the planet pushed and tugged on the many smaller continents that were home to generations of animals long before the dinosaurs, until all of the land was globbed together into one sprawling kingdom.

And what about the climate? No better way to put it: the earliest dinosaurs lived in a sauna. The Earth was a whole lot warmer back in the Triassic Period than it is today. In part, that's because there was more carbon dioxide in the atmosphere, so more of a greenhouse effect, more heat radiating across the land and sea. But the geography of Pangea exacerbated things. On one side of the globe, dry land extended from pole to pole, but on the other side, there was open ocean. That meant that currents could travel unimpeded from the equator to the poles, so there was a direct path for water baked in the low-latitude sun to heat up the high-latitude regions. This prevented ice caps from forming. Compared to today, the Arctic and Antarctic were balmy, with summer temperatures similar to those of London or San Francisco, and winter temperatures that barely inched below freezing. They were places that early dinosaurs and the other creatures with whom they shared the earth could easily inhabit.

If the poles were that warm, then the rest of the world must have been a hothouse. But it's not as though the entire planet was a desert. Once again the geography of Pangea made things much more complex. Because the supercontinent was basically centered on the equator, half the land was always scorching in the summer while the other half was cooling down in the winter. The marked temperature differences between north and south caused violent air currents to regularly stream across the equator. When the seasons changed, these currents shifted direction. That kind of thing happens today in some parts of the world, particularly India and Southeast Asia. It's what drives the monsoons, the alternation of a dry season with a prolonged deluge of rain and nasty storms. You've probably seen images in the newspaper or on the nightly news: floods drowning homes, people fleeing from raging torrents, mudslides burying villages. The modern monsoons are localized, but the Triassic ones were global. They were so severe that geologists have invented a hyperbolic term to describe them: megamonsoons.

Many a dinosaur was probably swept away by floodwaters or entombed by mud avalanches. But the megamonsoons also had another effect. They helped divide Pangea into environmental provinces, characterized by different amounts of precipitation, varying severity of the monsoonal winds, and different temperatures. The equatorial region was extremely hot and humid, a tropical hell that would make summer in today's Amazon seem a trip to Santa's workshop by comparison. Then there were vast stretches of desert, extending about 30 degrees of latitude on either side of the equator—like the Sahara, only covering a much broader swath of the planet. Temperatures here were well into the hundreds (over 35 degrees Celsius), probably all year long, and the monsoonal rains that pounded other parts of Pangea were absent here, offering little more than a trickle of precipitation. But the monsoons exerted a great impact in the midlatitudes. These areas were slightly cooler but much more wet and humid than the deserts, far more hospitable to life. *Herrerasaurus*, *Eoraptor*, and the other Ischigualasto dinosaurs lived in such a setting, smack in the middle of the midlatitude humid belt of southern Pangea.

Pangea may have been a united landmass, but its treacherous weather and extreme climates gave it a dangerous unpredictability. It wouldn't have been a particularly safe or pleasant place to call home. But the very first dinosaurs had no choice. They entered a world still recovering from the terrible mass extinction at the end of the Permian, a land subject to the violent whims of storms and the blight of blistering temperatures. So did many other new types of plants and animals that were getting their start after the mass extinction cleared the planet. All of these newbies were thrust onto an evolutionary battlefield. It was far from certain that dinosaurs were going to emerge triumphant. After all, they were small and meek creatures, nowhere near the top of the food chain during their earliest years. They were hanging around with lots of other species of small-to-midsize reptiles, early mammals, and amphibians in the middle of the food pyramid, fearful of the crocodile-line archosaurs, who held the throne. Nothing was handed to the dinosaurs. They were going to have to earn it.

DURING MANY SUMMERS, I've journeyed deep into the subtropical arid belt of northern Pangea, on the hunt for fossils. Of course, the supercontinent itself is long gone, having gradually fractured into our modern continents during the more than 230 million years since the primeval dinosaurs started their evolutionary march. What I've been exploring is a remnant of old

Pangea that can be found in the sunny Algarve region of Portugal, at the very southwestern corner of Europe. During those formative years when dinosaurs were navigating the megamonsoons and boiling heat waves of the Triassic, this part of Portugal was only 15 or 20 degrees north of the equator, about the same latitude as Central America today.

As with so many adventures in paleontology, it was a random clue that put Portugal on my radar. After our first jaunt together in Poland, visiting Grzegorz and studying fossils of some of the dinosauromorph ancestors of dinosaurs, my British buddy Richard Butler and I developed something of an addiction. We became obsessed with the Triassic Period. We wanted to understand what the world was like when dinosaurs were still young and vulnerable. So we scoured the map of Europe looking for other places where there were accessible rocks of Triassic age, the type of sediments that could conceivably contain the fossils of dinosaurs and other animals living alongside them. Richard came across a short paper in an obscure scientific journal, describing some scraps of bone from southern Portugal that were collected by a German geology student in the 1970s. The student had been in Portugal to make a map of the rock formations, a rite of passage for all undergraduate geology majors. He had little interest in fossils, so he threw the specimens in his rucksack and hauled them back to Berlin, where they languished in a museum for nearly three decades until some paleontologists recognized them as skull pieces of ancient amphibians. *Triassic* amphibians. That was enough to get us excited. There were Triassic fossils in a beautiful part of Europe and nobody had been looking for them for decades. We had to go.

That tip brought Richard and me to Portugal in the late summer of 2009, the hottest part of the year. We teamed up with another friend, Octávio Mateus, who wasn't even thirty-five years old at the time but was already regarded as Portugal's leading dinosaur hunter. Octavio grew up in a little town called Lourinhã, on the windy Atlantic coast north of Lisbon. His parents were amateur archaeologists and historians who spent weekends exploring the countryside, which just so happened to be strewn with Jurassic dinosaur fossils. The Mateus family and their ragtag band of local enthusiasts collected so many dinosaur bones, teeth, and eggs that they needed a place to put them, so when Octávio was nine years old, his parents started their own museum. Today, the Museu da Lourinhã houses one of the most important collections of dinosaurs in the world, many of which have been collected by Octávio—who went on to study paleontology and become a professor in Lisbon—and by his ever-expanding army of students, volunteers, and homegrown helpers.

It was fitting that Octavio, Richard, and I set out in the August heat, because we were chasing the fossils of animals that lived in the very hottest and driest sector of Pangea. But it wasn't very good strategy on our part. For several days, we hiked through the sun-baked hills of the Algarve, our sweat soaking the geological maps that we hoped would lead us to our treasure. We checked out nearly every speck of Triassic-age rock on the maps and relocated the site where the geology student had collected his amphibian bones, but all we saw were fossil crumbs. As our week in the field drew to a close, we were hot and exhausted, and staring down the barrel of failure. On the verge of defeat, we thought we should take one more hike in the area where the geology student made his discovery. It was a scorcher of a day, the thermometer on our handheld GPS units reaching 120 degrees Fahrenheit (50 Celsius).

After an hour or so of prospecting together, we decided to split up. I stayed near the base of the hills, scrutinizing the fragments of bone scattered across the ground in a desperate attempt to trace them to their source. I had no luck. But then I heard an excited voice scream from somewhere up on the ridge. I detected a hint of a lyrical Portuguese accent, so it must have been Octávio. I rushed toward where I thought the voice was coming from, but now there was nothing but silence. Maybe I was imagining things, the heat playing tricks on my brain. Eventually I saw Octávio in the distance, rubbing his eyes like someone woken up by a phone call in the middle of the night. He was stumbling, giving off a bit of a zombie vibe. It was weird.

When Octávio saw me, he gathered himself and burst into song. "I found it, I found it, I found it," he repeated over and over. He was holding a bone. What he didn't have was a water bottle. And suddenly it made sense. He had forgotten his water in the car, a bad thing for such a hot day, but he had happened upon the layer where the amphibian bones were coming out. The combination of exhilaration and dehydration had caused him to pass out for a moment. But now he was back into consciousness, and a few moments later, Richard had scrambled his way through the brush to join us. After exchanging excited hugs and high fives, we celebrated further by rehydrating with beers at a small café down the road.

What Octávio had found was a half-meter-thick layer of mudstone full of fossil bones. We returned several times over the next few years to meticulously excavate the site, which turned out to be a chore because the bone layer seemed to extend infinitely into the hillside. I had never seen so many fossils concentrated together in one area. It was a mass graveyard. Countless skeletons of amphibians called *Metoposaurus*—supersize versions of today's salamanders that were the size of a small car—were jumbled together in a chaotic mess. There must have been hundreds of them. Some 230 million years ago, a flock of these slimy, ugly monsters suddenly died when the lake they were living in dried up, collateral damage of the capricious Pangean climate.

Giant amphibians like *Metoposaurus* were leading actors in the story of Triassic Pangea. They prowled the shores of rivers and lakes over much of the supercontinent, particularly the subtropical arid regions and midlatitude humid belts. If you were a frail little primitive dinosaur like *Eoraptor*, you would want to avoid the shorelines at all costs. It was enemy territory. *Metoposaurus* was there waiting, lurking in the shallows, ready to ambush anything that ventured too close to the water. Its head was the size of a coffee table, and its jaws were studded with hundreds of piercing teeth. Its big, broad, almost flat upper and lower jaws were hinged together at the back and could snap shut like a toilet seat to gobble up whatever it wanted. It would only take a few bites to finish off a delicious dinosaur supper.

Salamanders bigger than humans seem like a mad hallucination. As bizarre as they were, though, *Metoposaurus* and its kin were not aliens. These terrifying predators were the ancestors of today's frogs, toads, newts, and salamanders. Their DNA flows through the veins of the frog hopping around your garden or the one you dissected in high school biology class. As a matter of fact, many of today's most recognizable animals can be traced back to the Triassic. The very first turtles, lizards, crocodiles, and even mammals came into the world during this

Excavating the *Metoposaurus* bone bed in Algarve, Portugal, with Octávio Mateus, Richard Butler, and our team.

time. All of these animals—so much a fabric of the Earth we call home today—rose up alongside the dinosaurs in the harsh surroundings of prehistoric Pangea. The apocalypse of the end-Permian extinction left such an empty playing field that there was space for all sorts of new creatures to evolve, which they did unabated during the 50 million years of the Triassic. It was a time of grand biological experimentation that changed the planet forever and reverberates still today. It's no wonder many paleontologists refer to the Triassic as the "dawn of the modern world."

If you could put yourself into the tiny feet of our furry, mouse-size Triassic mammalian ancestors, you would be looking up at a world that was starting to show whispers of today. Yes, the physical planet itself was completely different—a supercontinent, marked by intense heat and violent weather. Nevertheless, the parts of the land not engulfed by desert were covered in ferns and pine trees. There were lizards darting around in the forest canopy, turtles paddling in the rivers, amphibians running amok, many familiar types of insects buzzing around. And there were dinosaurs, mere bit characters in this ancient scene but destined for greater things to come.

AFTER SEVERAL YEARS of excavating the supersalamander mass grave in Portugal, we've collected a lot of bones of *Metoposaurus*, enough to fill the workshop in Octávio's museum. But we've also found other animals that died when the prehistoric lake evaporated. We dug up part of the skull of a phytosaur, a long-snouted relative of crocodiles that hunted on land and in the water. We've scooped up many teeth and bones of various fishes, which were probably the primary source of food for *Metoposaurus*. Other small bones hint at a badger-size reptile.

What we haven't found yet are any signs of dinosaurs.

It's strange. We know dinosaurs were living south of the equator, in the humid river valleys of Ischigualasto, at the same general time that *Metoposaurus* was terrorizing the lakes of Triassic Portugal. We also know that many different types of dinosaurs were commingling in Ischigualasto: all of those creatures that I studied in Ricardo Martínez's museum in Argentina. Meat-eating theropods like *Herrerasaurus* and *Eodromaeus*, primitive long-necked sauropod precursors like *Panphagia* and *Chromogisaurus*, early ornithischians (cousins of the horned and duck-billed dinosaurs). No, they weren't at the top of the food pyramid. Yes, they were outnumbered by the jumbo amphibians and crocodile relatives, but they were at least beginning to make their mark.

So why don't we see them in Portugal? It could be, of course, that we just haven't found them yet. Absence of evidence is not always evidence of absence, as all good paleontologists must continually remind themselves. Next time we go back into the scrublands of the Algarve and carve out another section of the bone bed, maybe we'll find ourselves a dinosaur. However, I'm willing to bet against that, because a pattern is starting to emerge as paleontologists discover more and more Triassic fossils from around the world. Dinosaurs seem to be present and starting to slowly diversify in the temperate humid parts of Pangea, particularly in the southern hemisphere, during a slice of time from about 230 to 220 million years ago. Not only do we find their fossils in Ischigualasto, but also in parts of Brazil and India that were once in the Pangean humid zone. Meanwhile, in the arid belts closer to the equator, dinosaurs were absent or extremely rare. Just as in Portugal, there are great fossil sites in Spain, Morocco, and along the eastern coast of North America where you can find plenty of amphibians and reptiles, but nary a dinosaur. All of these places were in the parched arid sector of Pangea during those 10 million years when dinosaurs were beginning to blossom in the more bearable humid regions. It seems these first dinosaurs couldn't handle the desert heat.

It's an unexpected story line. Dinosaurs didn't just sweep across Pangea the moment they originated, like some infectious virus. They were geographically localized, held in place not by physical barricades but by climates they couldn't endure. For many millions of years, it looked as if they might remain provincial rubes, stuck in one zone in the south of the supercontinent, unable to break free—an aging high school football hero of faded dreams, who could have been something if only he'd been able to get out of his tiny hometown.

Underdogs—that's what these first humidity-loving dinosaurs were. They wouldn't have been a very impressive bunch. Not only were they trapped by the deserts, but even where they were able to eke out a living, they were barely getting by, at least at first. True, there were several dinosaur species in Ischigualasto, but these made up only about 10 to 20 percent of the total ecosystem. They were vastly outnumbered by early mammal relatives, like the pig-mimic dicynodonts that ate roots and leaves, and by other types of reptiles, most notably rhynchosaurs, which chopped plants with their sharp beaks, and crocodile cousins like the mighty apex predator *Saurosuchus*. At the same time but slightly to the east, in what is now Brazil, the story was much the same. There were a few different types of dinosaurs closely related to species in Ischigualasto: the carnivorous *Staurikosaurus* was a cousin of *Herrerasaurus*, and the small long-necked creature *Saturnalia* was very similar to *Panphagia*. But they were quite rare, again overwhelmed by masses of proto-mammals and rhynchosaurs. Even farther to the east, where the humid zone continued into what is now India, there were a handful of primitive long-necked sauropod relatives, like *Nambalia* and *Jaklapallisaurus*, but once again they were role players in ecosystems ruled by other species.

Then, when it appeared that dinosaurs were never going to

escape their rut, two important things happened that gave them an opening.

First, in the humid belt, the dominant large plant-eaters, the rhynchosaurs and dicynodonts, became less common. In some areas they disappeared entirely. We don't yet fully understand why, but the consequences were unmistakable. The fall of these herbivores gave the plant-eating primitive sauropod cousins like *Panphagia* and *Saturnalia* an opportunity to seize a new niche in some ecosystems. Before long they were the main herbivores in the humid regions of both the southern and northern hemispheres. In the Los Colorados Formation of Argentina, a unit of rock laid down from about 225 to 215 million years ago that was formed directly after the Ischigualasto dinosaurs left their fossils, the sauropod antecedents are the most common vertebrates. There are more fossils of these cow-to-giraffe-size plant-guzzlers—among them *Lessemsaurus*, *Riojasaurus*, and *Coloradisaurus*—than any other type of animal. In all, dinosaurs comprise about 30 percent of the ecosystem, while the once dominant mammal relatives dip below 20 percent.

It wasn't only a southern Pangean story. Across the equator in primeval Europe, then part of the Northern Hemisphere humid sector, other long-necked dinosaurs were also thriving. As in Los Colorados, they were the most common large plant-eaters in their habitats. One of these species, *Plateosaurus*, has been found at over fifty sites throughout Germany, Switzerland, and France. There are even mass graves like the *Metoposaurus* bone bed in Portugal, where dozens (or more) of *Plateosaurus*es died together when the weather turned rough, a sign of just how many of these dinosaurs were flocking across the landscape.

The second major breakthrough, around 215 million years

ago, was that the first dinosaurs began arriving in the subtropical arid environments of the Northern Hemisphere, then about 10 degrees above the equator, now part of the American Southwest. We don't know exactly why dinosaurs were now able to migrate out of their safe humid homes and into the harsh deserts. It probably had something to do with climate change—shifts in the monsoons and the amount of carbon dioxide in the atmosphere made the differences between the humid and arid regions less stark, so dinosaurs could move more easily between them. Whatever the reason, at long last dinosaurs were making inroads into the tropics, expanding into parts of the world that had previously eluded them.

The best records of desert-living Triassic dinosaurs come from areas that are once again deserts today. Across much of the postcard-pretty landscape of northern Arizona and New Mexico are hoodoos, badlands, and canyons carved out of colorful red and purple rocks. These are the sandstones and mudstones of the Chinle Formation, a third-of-a-mile-thick rock sequence formed from the ancient sand dunes and oases of tropical Pangea during the last half of the Triassic, from about 225 to 200 million years ago. Petrified Forest National Park, which should be on the itinerary of any dino-loving tourist visiting the southwestern states, has one of the best exposures of the Chinle Formation, full of thousands of enormous fossilized trees that were uprooted and buried in flash floods right around the time that dinosaurs were starting to settle in the area.

Some of the most exciting paleontological fieldwork over the past decade has targeted the Chinle Formation. New discoveries have painted a striking new image of what the first desert-dwelling dinosaurs were like and how they fit into their broader ecosystems. Leading the charge is a remarkable group of young researchers, who were all graduate students when they began exploring the Chinle. The core of the group is the four-man band of Randy Irmis, Sterling Nesbitt, Nate Smith, and Alan Turner. Irmis is a bespectacled introvert but a beast of a field geologist; Nesbitt is an expert on fossil anatomy who's always wearing a baseball cap and quoting television comedy shows; Smith is a smooth-dressing Chicagoan who likes to use statistics to study dinosaur evolution; and Turner, an expert on building family trees of extinct groups, is affectionately called Little Jesus because of his flowing locks, bushy beard, and moderate stature.

The quartet is a half generation ahead of me on the career path. They were working on their PhDs when I was starting to do research as an undergraduate. As a young student, I was in awe of them, as if they were a paleontology Rat Pack. They traveled in a herd at research conferences, often with other friends of theirs who worked in the Chinle: Sarah Werning, a specialist on how dinosaurs and other reptiles grew; Jessica Whiteside, a brilliant geologist who studied mass extinctions and ecosystem changes in deep time; Bill Parker, the paleontologist at Petrified Forest National Park and an expert on some of the close crocodile relatives who lived with early dinosaurs; Michelle Stocker, who studied some of the other proto-crocodiles (and whom Sterling Nesbitt later convinced to marry him—proposing on a field trip, no less—forming a different sort of Triassic dream team). They were the hotshot young scientists whom I looked up to, the type of researchers I wanted to become.

For many years, the Chinle Rat Pack has been spending summers in northern New Mexico, in the pastel drylands near the tiny hamlet of Abiquiú. In the mid-1800s, this outpost was an important stop on the Old Spanish Trail, a trade route that linked nearby Santa Fe with Los Angeles. Today only a few hundred people remain, making the area feel like a remote backwater within the world's most industrialized country. Some people like that kind of seclusion, though. One of them was Georgia O'Keeffe, the modernist American artist famous for her paintings of flowers that were intimate to the point of abstraction. O'Keeffe was also drawn to sweeping landscapes, and she was moved by the striking beauty and incomparable hues of natural light in the Abiquiú area. She bought a house nearby, on the sprawling grounds of a desert retreat called Ghost Ranch. There she could explore nature and experiment with new painting styles without being bothered by anyone. The red cliffs and colorful candy-striped canyons of the ranch, bathed in sparkling sunbursts, are common motifs in the work she produced here.

After O'Keeffe died, in the mid-1980s, Ghost Ranch became a pilgrimage site for art lovers hoping to catch some of that desert spark that so inspired the old master. Few of these cultured travelers probably realize that Ghost Ranch is also bursting with dinosaur bones.

But the Rat Pack knew.

They understood that in 1881 a scientific mercenary named David Baldwin had been sent to northern New Mexico by the Philadelphia paleontologist Edward Drinker Cope, with the singular mission to find fossils that Cope could stick in the face of his Yale rival, Othniel Charles Marsh. The two Easterners were engaged in a bitter feud known to history as the Bone Wars (of which, more later), but by this stage of their careers, neither of them particularly liked to brave the elements and Native

American war parties—Geronimo would continue raiding New Mexico and Arizona until 1886. Rather than look for fossils themselves, they relied on a network of hired guns. Baldwin was the type of character they often employed: a mysterious loner who would jump on his mule and head deep into the badlands for months at a time, even during the bleak winter, and eventually emerge loaded up with dinosaur bones. In fact, Baldwin had worked for both of the pugnacious paleontologists: he was once a trusted confidant of Marsh's, but now his loyalties were with Cope. Thus it was Cope who was the lucky recipient of the collection of small, hollow dinosaur bones that Baldwin pried out of the desert near Ghost Ranch. These bones belonged to a totally new type of dog-size, lightweight, fast-running, sharp-toothed, primitive Triassic dinosaur Cope later called *Coelophysis*. Like *Herrerasaurus* from Argentina, which would be found many decades later, it was one of the earliest members of the theropod dynasty that would eventually produce *T. rex*, *Velociraptor*, and birds.

The Chinle Rat Pack also knew that a half century after Baldwin's discovery, another East Coast paleontologist, Edwin Colbert, took a liking to the Ghost Ranch area. He was a much more pleasant individual than Cope or Marsh. When Colbert set out for Ghost Ranch in 1947, he was in his early forties, already ensconced in one of the top jobs in the field: curator of vertebrate paleontology at New York City's American Museum of Natural History. That summer, while O'Keeffe was painting mesas and rock sculptures only a few miles away, Colbert's field assistant George Whitaker made an astounding discovery. He came across a *Coelophysis* graveyard, hundreds of skeletons in all, a pack of predators buried by a freak flood. I can imagine he must have felt something similar to our unbridled joy when we found our *Metoposaurus* bone bed in Portugal. Overnight *Coelophysis* became the quintessential Triassic dinosaur, the creature that immediately came to mind when people envisioned what the earliest dinosaurs looked like, how they behaved, and what environments they lived in. For years the American Museum crew kept digging and digging, hacking out blocks of the bone bed, which were distributed to museums around the world. Odds are, if you go see a big dinosaur exhibit today, you'll see a Ghost Ranch *Coelophysis*.

The Chinle Rat Pack also knew of one final, and perhaps most important, clue. Because so many *Coelophysis* skeletons were found together, excavating the mass grave site diverted everyone's attention for decades. It sucked up most of the money for fieldwork, most of the time and energy of the field crews. But it was merely a single site in the expanse of Ghost Ranch, tens of thousands of acres covered by fossil-rich Chinle rocks. More must have been out there. So it was no surprise to them when, in 2002, a retired forest manager named John Hayden discovered some bones while hiking less than half a mile from the main gate of Ghost Ranch.

A few years later, the team of Irmis, Nesbitt, Smith, and Turner returned to the spot, got out their tools, and started digging. It took a lot of time and a lot of sweat. Once, when I was catching up with the quartet in a New York City Irish pub, Nate Smith turned to me, cocked his head up toward the ceiling, and said with a hint of cheeky machismo, "The amount of rock we removed that summer, yeah, it would fill up this bar."

But the toil was worth it. The crew confirmed that there were indeed fossil bones at the site. Then they kept finding more and

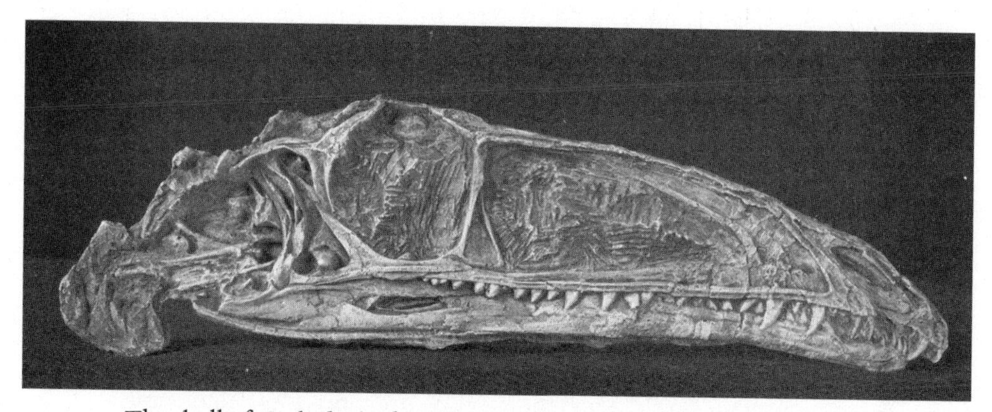

The skull of *Coelophysis*, the primitive theropod found in abundance at Ghost Ranch. *Courtesy of Larry Witmer.*

more of them, hundreds, thousands. It turned out to be a river channel deposit, where currents had dumped the skeletons of many unlucky creatures swept into the water some 212 million years ago. With the right cocktail of good detective work and a drive to make their own discoveries even though they were still students, the Rat Pack had unearthed a treasure trove of Triassic fossils. The site—nicknamed the Hayden Quarry after the sharp-eyed forester who noticed the first fossil eroding out of the ground—has become one of the world's most important Triassic fossil localities.

The quarry provides a snapshot of an ancient ecosystem, one of the first deserts that dinosaurs were able to live in. It wasn't the picture the Chinle Rat Pack was expecting. When the young mavericks started digging in the mid-2000s, the prevailing wisdom was that dinosaurs conquered the deserts soon after they arrived in the Late Triassic. Other scientists had collected a wealth of fossils from similar-age rock units in New Mexico, Arizona, and Texas, which seemed to belong to more than a dozen species of dinosaurs, ranging from stocky apex predators and smaller meat-eaters to many different types of plant-munching ornithischians, the ancestors of *Triceratops* and the duckbills. It seemed that dinosaurs were everywhere. But that wasn't the case in the Hayden Quarry. There were monster amphibians closely related to our Portuguese *Metoposaurus*, primitive crocodiles and some of their long-snouted and armored relatives, skinny reptiles with short legs called *Vancleavea*, which looked like scaly dachshunds, and even funny little reptiles that hung from the trees like chameleons, called drepanosaurs. Those are the common animals in the quarry. Dinosaurs were anything but. The Rat Pack found only three types of dinosaurs: a fleet-footed predator very similar to Baldwin's *Coelophysis*, another swift carnivore called *Tawa*, and a somewhat larger and stockier meat-eater called *Chindesaurus*, which was closely related to the Argentine *Herrerasaurus*. Each is represented by only a few fossils.

It was a great surprise to the team. Dinosaurs were rare in the tropical deserts of the Late Triassic, and it was only the meat-eaters that seemed to be hanging about. There were no plant-eating dinosaurs, none of the ancestral long-necked species that were so common in the humid zones, none of the ornithischian forebears of *Triceratops*. It's a meager bunch of dinosaurs surrounded by all sorts of bigger, meaner, more common, more diverse animals.

What, then, to make of the dozens of Triassic dinosaur species that other scientists had identified from all over the American Southwest? Irmis, Nesbitt, Smith, and Turner scrutinized all of the evidence they could find, traveling to every small-town museum where researchers had deposited their fossils. They saw that most of these specimens were isolated teeth and scraps of bone, not the best foundation for naming new species. But that wasn't the shocker. The more they found at Hayden Quarry, the better search image the crew developed in their heads. They became able to tell a dinosaur from a crocodile from an amphibian almost by instinct. In a series of eureka moments, they realized that most of those supposed dinosaur fossils collected by others weren't dinosaurs at all, but primitive dinosauromorph cousins of dinosaurs or, in some cases, early crocodiles and their kin that just so happened to look like dinosaurs.

So not only were dinosaurs rare in the Late Triassic deserts, but they were still living alongside their archaic relatives, the same types of animals that were leaving their tiny footprints in Poland nearly 40 million years earlier. It was a jarring realization. Up until then, almost everyone thought that the primitive dinosauromorphs were an uninteresting ancestral stock whose only destiny was to give birth to the mighty dinosaurs. Once that job was done, they could quietly fade away to extinction. But here they were, all over Late Triassic North America, even a new poodle-size species called *Dromomeron* in the Hayden Quarry, living alongside proper dinosaurs for some 20 million years.

Probably the only person not surprised by the findings was another student, an Argentine named Martín Ezcurra. Independently of the four American grad students, Martín was starting to doubt the identifications of some of the supposed North American "dinosaurs" collected by the older generation of paleontologists, but he didn't have the resources to go study them, because he was from South America and still learning English.

That, and he was a teenager.

One thing he did have, however, was access to the tremendous collections of Ischigualasto dinosaurs from his home country, thanks to the generosity of Ricardo Martínez and other curators who responded positively to the unusual request of a high schooler wanting to visit their museums. Martín gathered photos of many of the mysterious North American specimens and carefully compared them to the Argentine dinosaurs, and recognized that there were key differences. One North American species in particular, a skinny carnivore called *Eucoelophysis*, which was supposed to be a theropod, was actually a primitive dinosauromorph. He published this result in a scientific journal in 2006, the year before Irmis, Nesbitt, Smith, and Turner published their first findings. Martín was seventeen years old when he wrote his paper.

It's hard to fathom why dinosaurs were doing so poorly in the deserts while so many other animals, including their dinosauromorph precursors, were having a better go at it. To get to the bottom of the question, Chinle's Rat Pack collaborated with the skilled geologist Jessica Whiteside, who was also part of our excavation teams in Portugal. Jessica is a maestro at reading the rocks. Better than anyone I've ever known, she can look at a sequence of rocks and tell you how old they are, what the environments were like when they formed, how hot it was, even how much rain there was. Set her loose at a fossil site, and she'll come back with a story from the distant past of changing climates, shifting weather, evolutionary explosions, and great extinctions.

Jessica put her sixth sense to use at Ghost Ranch and determined that the animals of the Hayden Quarry did not have an easy life. They lived in an environment that wasn't always a desert, but one in which seasonal climates dramatically fluctuated.

It was bone-dry for much of the year, but wetter and cooler during other times—hyperseasonality, as Jessica and the Rat Pack call it. The culprit was carbon dioxide. Jessica's measurements show that there were somewhere around 2,500 molecules of carbon dioxide per every million molecules of air in the tropical regions of Pangea back when the Hayden Quarry animals were alive. That's more than six times the amount of carbon dioxide today. Let that sink in for a minute—just think about how quickly temperatures are rising now and how anxious we are about future climate change, even though there is much less carbon dioxide in today's atmosphere. The high concentration of carbon dioxide in the Late Triassic started a chain reaction: huge fluctuations in temperature and precipitation, raging wildfires during parts of the year but humid spells in others. Stable plant communities had a difficult time establishing themselves.

It was a chaotic, unpredictable, unstable part of Pangea. Some animals could deal with that better than others. Dinosaurs seem to have been able to cope a little bit, but not able to truly thrive. The smaller meat-eating theropods were able to manage, but the larger, fast-growing plant-eaters, which required a steadier diet, could not. Even some 20 million years after they had originated, even after they had taken over the large-herbivore niche in humid ecosystems and started to colonize the hotter tropics, dinosaurs were still having trouble with the weather.

IF YOU WERE standing on safe ground during a Late Triassic flood, watching the animals eventually buried at Hayden Quarry get swept up by the seasonal river that drowned them, you might have had a hard time telling some of the corpses apart as they floated by. Sure, it would be easy to recognize one of the giant supersalamanders or some of those weird chameleon-mimic reptiles. But you might not be able to distinguish dinosaurs like *Coelophysis* and *Chindesaurus* from some of the crocodiles and their kin. Even if you were able to watch these animals alive, going about their business of eating and moving and interacting with each other, you still might have trouble.

Why the confusion? It's the same reason that the previous generation of paleontologists working in the American Southwest so often misidentified crocodile fossils as dinosaurs, and why other scientists in Europe and South America made the same mistakes. During the Late Triassic, there were many other animals that really, really looked and behaved like dinosaurs. In evolutionary biology speak, this is called convergence: different types of creatures resembling each other because of similarities in lifestyle and environment. It's why birds and bats, which both fly, each have wings. It's why snakes and worms, which both squirm through underground burrows, are both long, skinny, and legless.

The convergence between dinosaurs and crocodiles is surprising, shocking even. The alligators that prowl the Mississippi delta and the crocodiles that lurk in the Nile may appear vaguely prehistoric, but they don't look anything like a *T. rex* or a *Brontosaurus*. During the Late Triassic, however, crocodiles were very different.

Recall that dinosaurs and crocodiles are both archosaurs—members of that large group of upright-walking reptiles that started to blossom after the end-Permian mass extinction, which proliferated because they could move much faster and more efficiently than the sprawling animals of the time. Early in the Triassic, archosaurs split into two major clans: the avemetatarsalians, which led to dinosauromorphs and dinosaurs, and the pseudosuchians, which gave rise to crocodiles. During the dizzying splurge of postextinction evolution, the pseudosuchian tribe also produced a number of other subgroups that diversified in the Triassic but then went extinct. Because they don't survive today—unlike the crocodiles and dinosaurs (in the guise of birds)—these groups have largely been forgotten about, considered oddities from a distant past, evolutionary dead ends that never rose to the top. That stereotype is wrong, though, because for much of the Triassic these crocodile-line archosaurs were thriving.

Most of the major types of Late Triassic pseudosuchians can be found at Hayden Quarry. There is a phytosaur called *Machaeroprosopus*, a member of that group of long-snouted, semiaquatic ambush predators whose bones we also found in Portugal. They were bigger than a motorboat and snatched fish—and the occasional passing dinosaur—with the hundreds of spiky teeth in their stretched jaws. It was neighbors with *Typothorax*, a plant-eater built like a tank with armor covering its body and spikes sticking out from its neck. It belongs to a group called the aetosaurs, a hugely successful family of mid-tier herbivores that closely resembled the armored ankylosaur dinosaurs that evolved millions of years later. They were good diggers and may have even cared for their young by building and guarding nests. Then there are proper crocodiles, but nothing like the ones we're familiar with today. These primitive Triassic species—the ancestral breed that modern crocs evolved from—looked like greyhounds: they were about the same size, stood on four legs, had the emaciated build of a supermodel, and could sprint like champions. They fed on bugs and lizards and were most certainly not top predators. That title went to the rauisuchians, a ferocious bunch that grew up to twenty-five feet long, bigger than the largest saltwater crocodiles today. We met one of them previously, *Saurosuchus*, the top gun in the Ischigualasto ecosystem that would have haunted the nightmares of the very first dinosaurs. Imagine a slightly smaller version of a *T. rex* walking around on four legs, with a muscular skull and neck, railroad-spike teeth, and a bone-breaking bite.

There's also another type of crocodile-line archosaur found at Ghost Ranch—not in the Hayden Quarry itself, but in the nearby *Coelophysis* graveyard. It was found in 1947, not long after Whitaker discovered the bone bed, during those first few weeks of excavation. The American Museum team was digging up so many *Coelophysis* skeletons that, after a while, the excitement wore off and they got a little bored. Everything they saw started to look like *Coelophysis*. So they didn't notice that one of the skeletons they collected was similar in size to *Coelophysis*, and had the same long legs and light build, but was a little different in other ways—notably, it had a beak instead of an arsenal of sharp teeth. The technicians back in New York didn't notice either. They started to remove the specimen from the block of rock it was embedded in, but were all too keen to stop once they determined it was just another *Coelophysis*. It could go in the storehouse with the rest of them.

The fossil stayed in the bowels of the museum, unconserved and unloved, until 2004. That's when one of the Ghost Ranch quartet, Sterling Nesbitt, started his PhD at Columbia University in New York. Because he was planning a project on Triassic dinosaurs, he went back through all of the fossils collected by

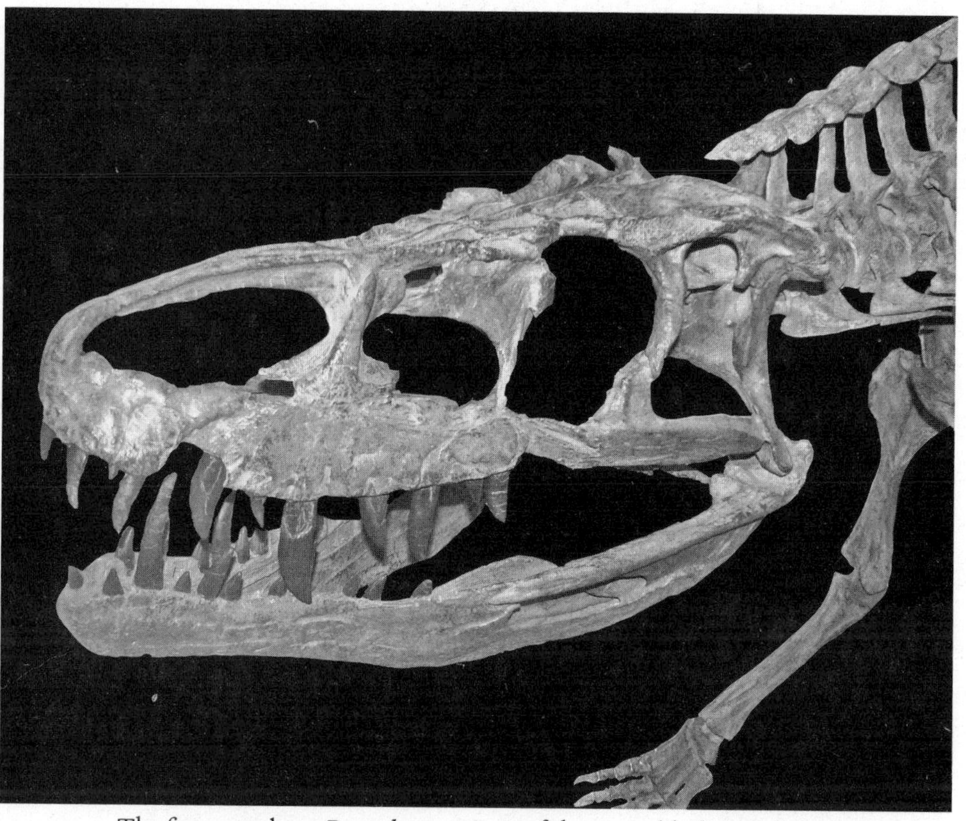

The fierce predator *Batrachotomus*, one of the crocodile-line archosaurs (rauisuchians) that preyed on early dinosaurs.

Colbert, Whitaker, and their teams in the 1940s. Many were still encased in plaster, so they would have to remain on the shelves. But that one block from 1947 had been opened and partially prepared by the conservators, so Sterling could study it. With an excited pair of eyes and an enthusiasm that escaped the weary field hands a half century earlier, Sterling recognized that he wasn't looking at any old *Coelophysis*. He saw that it had a beak; he realized that its body proportions were different, that its arms were tiny. And then he noticed features of the ankle that were nearly identical to those of crocodiles. He wasn't looking at a dinosaur at all; he was looking at a pseudosuchian that was heavily convergent on dinosaurs.

This was the sort of discovery that young scientists dream about when secluded, alone with their thoughts, trawling through the drawers of museum collections. Since Sterling discovered it, he got to name it, and he chose the evocative moniker *Effigia okeeffeae*: the first name being the Latin word for ghost, in reference to Ghost Ranch, and the second paying homage to the ranch's most famous resident. *Effigia* made international headlines: the media loved this awkward-looking, toothless, stub-armed ancient crocodilian creature trying to pretend that it was a dinosaur. Stephen Colbert even devoted a segment of his show to the new discovery, complaining in jest that it should have been named after Edwin Colbert (who coincidentally shared a surname with the comedian) and not the feminist artist. I remember seeing that segment during the last year of my undergraduate studies, right around the time I was starting to plot out my own graduate-school future, and being in awe that a young grad student's work could make such an impact.

It also motivated me. Up until that point, I had been studying only dinosaurs, but I started to grasp that *Effigia* and the other dino-imitating pseudosuchians were critical in understanding how dinosaurs ascended to power. I started to read many of the classic studies in dinosaur paleontology, works by giants like Robert Bakker and Alan Charig, which were effusive in arguing that dinosaurs were special. They were so well endowed with superior speed, agility, metabolism, and intelligence that they outcompeted all of the other Triassic animals—the giant salamanders, the early mammal-like synapsids, and the crocodile-line pseudosuchians. Dinosaurs were the chosen ones. It was their manifest destiny to take on the weaker species, best them, and establish a global empire. There was almost a religious feel to some of these writings, perhaps not a surprise, given that Bakker also dabbles as an ecumenical Christian preacher and is renowned for his high-energy lectures, delivered in the style of an evangelist testifying to his congregation.

Dinosaurs outmaneuvering their foes on the Late Triassic battlefield. It was a good story, but it didn't sit well with me. New discoveries seemed to be upending the narrative, and a lot of that had to do with the pseudosuchians. So many of these crocodile-line archosaurs were dead ringers for dinosaurs. Or maybe it was the other way around: maybe Triassic dinosaurs were trying to be pseudosuchians. Regardless, if the two groups were similar in so many ways, then how could you argue that dinosaurs were a superior race? And it wasn't only the convergence between dinosaurs and pseudosuchians that threw up a red flag. There were *more* pseudosuchians than dinosaurs in the Late Triassic: more species and a greater abundance of these species in individual ecosystems. The menagerie of croc cousins from Ghost Ranch—phytosaurs, aetosaurs, rauisuchians, *Effigia*-like animals, true crocodiles—was not a local phenomenon. These were diverse groups that prospered throughout much of the world.

But, as scientists often like to say when trying to critique each other with subtlety, this all sounded a little arm-wavy. Could we somehow compare explicitly how dinosaurs and pseudosuchians were evolving in the Late Triassic? Was there a way to test whether one group was more successful than the other and whether that was changing over time? I buried myself in literature on statistics, unfamiliar territory for somebody who was consumed by dinosaurs but not yet very aware of other fields and techniques. I was a bit embarrassed to realize that invertebrate paleontologists—our redheaded stepsiblings, who study fossils like clams and corals, which don't have bones—had come up with a method two decades earlier, one that had been ignored by dinosaur workers. It was something called morphological disparity.

Morphological disparity sounds like a fancy term, but it is simply a measure of diversity. You can measure diversity in many ways. Counting up the number of species is one tack: you can say that South America is more diverse than Europe because there are more animal species there. Or you can compute diversity based on abundance: insects are more diverse than mammals because there are more insects in any given ecosystem. What morphological disparity does is measure diversity based on features of the anatomy. Thinking this way, you can consider birds to be more diverse than jellyfish, because birds have a much more complex body with lots of different parts, whereas jellyfish are just sacs of goo. This type of diversity measure can give great insight into evolution, because so many aspects of animal biology, behavior, diet, growth, and metabolism are controlled by anatomy. If you really want to know how a group is changing over time or how two groups compare in diversity, I would argue that morphological disparity is the most powerful way to do so.

Counting the number of species or the abundance of individuals is easy. All you need are a good set of eyes and a calculator. But how to measure morphological disparity? How to take all the complexity of the animal body and turn it into a statistic? I followed the approach pioneered by the invertebrate paleontologists. It went something like this. I first came up with a list of all of the Triassic dinosaurs and pseudosuchians, as these were the animals that I wanted to compare. I then spent months studying the fossils of these species and made a list of hundreds of features of the skeleton in which they vary. Some have five toes, others have three. Some walk on four legs, others on two. Some have teeth, others do not. I encoded these features in a spreadsheet as zeros and ones, just as a computer programmer would. *Herrerasaurus* walks on two legs, state 0. *Saurosuchus* walks on four legs, state 1. At the end of nearly a year of work, I had a database with seventy-six Triassic species, each assessed for 470 features of the skeleton.

With the long slog of data collection done, it was time for the math. The next step was to make what is called a distance matrix. It quantifies how different each species is from every other species, based on the database of anatomical characteristics. If two species share all features, then their distance score is 0. They are identical. If two other species share no characteristics, then their distance score is 1. They are completely different. For the in-between cases, let's say that *Herrerasaurus* and *Saurosuchus* share 100 characteristics but differ in the 370 others. Their distance score would be 0.79: the 370 features they differ in divided by the 470 total features in the data set. The best way to envision this is to think of those tables in a road atlas, which give distances between different cities. Chicago is 180 miles from Indianapolis. Indianapolis is 1,700 miles from Phoenix. Phoenix is 1,800 miles from Chicago. That table is a distance matrix.

Here's the neat trick about a distance matrix in an atlas. You can take that table of road distances between cities, stick it into a statistics software program, run what is called a multivariate analysis, and the program will spit out a plot. Each city will be a point on that plot, and the points will be separated by distance, in perfect proportion. In other words, the plot is a map—a geographically correct map with all of the cities in the right places and distances relative to each other. So what happens if we instead input the distance matrix that encapsulates the skeletal differences of Triassic dinosaurs and pseudosuchians? The statistics program will also produce a plot in which each species is represented by a point, a plot that scientists call a morphospace. But really it is just a map. It visually shows the spread of anatomical diversity among the animals in question. Two species close together have very similar skeletons, just as Chicago and Indianapolis are comparatively near geographically. Two species at far corners of the graph have very different anatomies, like the longer distance between Chicago and Phoenix.

This map of Triassic dinosaurs and pseudosuchians allows us to measure morphological disparity. We can group the animals in the plot by which great tribe they belong to—dinosaurs or pseudosuchians—and calculate which is occupying a larger swath of that map and therefore more anatomically diverse. In the same vein, we can further group the animals by time—Middle Triassic versus Late Triassic, let's say—and see if dinosaurs or pseudosuchians were becoming more or less anatomically diverse as the Triassic progressed. We did that and came up with a startling result that we published in 2008 in a study that helped launch my career. All throughout the Triassic, the pseudosuchians were significantly more morphologically diverse than dinosaurs. They filled a larger spread of that map, meaning they had a greater range of anatomical features, which indicated that they were experimenting with more diets, more behaviors, more ways of making a living. Both groups were becoming more diverse as the Triassic unfolded, but the pseudosuchians were always outpacing the dinosaurs. Far from being superior warriors slaying their competitors, dinosaurs were being overshadowed by their crocodile-line rivals during the 30 million years they coexisted in the Triassic.

PUT YOURSELF BACK in the tiny furry feet of our Triassic mammalian ancestors, surveying the Pangean scene as the Triassic was drawing to a close 201 million years ago. You would be seeing dinosaurs, but you wouldn't be surrounded by them. Depending on where you were, you might not have noticed them at all. They were relatively diverse in the humid regions, where protosauropods got as large as giraffes and were the most abundant plant-eaters, but there carnivorous theropods and herbivorous to omnivorous ornithischians were considerably smaller and less common. In the more arid zones, there were only small meat-eaters, the herbivores and larger species being unable to tolerate the hyperseasonal weather and megamonsoons. There were no dinosaurs remotely approaching a *Brontosaurus* or *T. rex* in size, and all across the supercontinent they were living under the thumb of their much more diverse, much more successful pseudosuchian adversaries. You would probably consider the dinosaurs a fairly marginal group. They were doing OK, but so were many other newly evolved types of animals. If you were of a gambling persuasion, you would probably have bet on one of these other groups, most likely those pesky crocodile-line archosaurs, as the ones that would eventually become dominant, grow to massive sizes, and conquer the world.

Some 30 million years after they originated, the dinosaurs had yet to mount a global revolution.

3

DINOSAURS BECOME DOMINANT

Scottish sauropod

SOME TIME AROUND 240 MILLION YEARS ago, the Earth began to crack. True dinosaurs hadn't quite evolved yet, but their cat-size dinosauromorph ancestors would have been there to experience the cracking—except there wasn't much to experience, at least not yet. There may have been some minor earthquakes, but these probably wouldn't have even registered with the dinosauromorphs, who were busy with more important things like fending off the supersalamanders and surviving the megamonsoons. As these dinosauromorphs gave rise to dinosaurs, the fracturing continued, many thousands of feet underground. Imperceptible on the surface, these fissures were slowly moving, growing, merging together, a hidden danger lurking under the feet of *Herrerasaurus*, *Eoraptor*, and the other first dinosaurs.

The very foundation of Pangea was splitting, and with the blissful ignorance of homeowners who don't realize there's a creeping crack in their basement until their house comes tumbling down, the dinosaurs had no inkling that their world was going to dramatically change.

As these earliest dinosaurs were evolving in fits and starts during the final 30 million years of the Triassic, great geological forces were tugging on Pangea from both the east and west. These forces—a planet-scale cocktail of gravity, heat, and pressure—are strong enough to make continents move over time. Because the pull was coming from two opposite directions, Pangea began to stretch and gradually become thinner, each small earthquake causing another tear. Imagine Pangea as a giant pizza, being torn apart by two hungry friends at opposite ends of the table: the crust becomes thinner until there is a rupture and it breaks into two. The same thing happened with the supercontinent. After a few tens of millions of years of slow and steady tug-of-war, east versus west, the cracks reached the surface, and the giant landmass began to unzip down its middle.

It's because of that ancient divorce between east and west Pangea that the seaboard of North America is separated from western Europe and South America sits apart from Africa. It's why there is now an Atlantic Ocean, which didn't exist until seawater rushed in to fill the gap between the separating tracts of land. Those forces and fractures over 200 million years ago shaped our modern geography. But there was more to it than that, because continents don't just split up and call it a day. As with human relationships, things can get really nasty when a continent breaks up. And the dinosaurs and other animals growing up on Pangea were about to be changed forever by the aftereffects of their home being ripped in two.

The problem boils down to this: as a continent tears, it bleeds lava. It's nothing more than basic physics. The Earth's outer crust is pulled apart and thins, decreasing pressure on the deeper parts of the Earth. As pressure lessens, magma from the deeper Earth rises to the surface and erupts through volcanoes. If there is only a little rip in the crust—two small bits of a continent separating from each other, let's say—then the effects aren't too bad. You might get a few volcanoes, some lava and ash, some local destruction, and then eventually it stops. That kind of thing is happening in eastern Africa today, and it's far from catastrophic. But if you're slashing apart an entire supercontinent, then you approach the realm of apocalypse.

At the very end of the Triassic, 201 million years ago, the world was violently remade. For 40 million years, Pangea had been gradually splintering apart, and magma had been welling underground. Now that the supercontinent had finally cracked, the magma had somewhere to go. Like a hot-air balloon rising through the sky, the liquid-rock reservoir rushed upward, broke through the shattered surface of Pangea, and gushed out onto the land. As with the volcanoes that had erupted at the end of the Permian Period some 50 million years earlier, causing the extinction that allowed dinosaurs and their archosaur cousins to get their start, these end-Triassic eruptions were different from any that humans have ever witnessed. We're not talking Pinatubo here, with hot clouds of ash bursting into the sky. Instead, over a period of some six hundred thousand years, there were four big pulses of drama, when enormous amounts of lava would surge out of the Pangean rift zone like tsunamis from hell. I'm hardly exaggerating: some of the flows were, added up together, up to three thousand feet thick; they could have buried the Empire State Building twice over. In all, some three million square miles of central Pangea were drowned in lava.

It goes without saying that this was a bad time to be a dinosaur, or for that matter, any other type of animal. These were some of the largest volcanic eruptions in Earth history. Not only did lava smother the land, but noxious gases that rode up with the lava poisoned the atmosphere and caused runaway global warming. These things triggered one of the biggest mass extinctions in the history of life, a mass die-off that claimed over 30 percent of all species and maybe much more. Paradoxically, however, it was also a mass extinction that helped dinosaurs break out of their early-life slump and become the enormous, dominant animals that stoke our imaginations.

IF YOU'RE WALKING down Broadway in New York City and happen to catch a gap between the skyscrapers, you can see straight across the Hudson River to New Jersey. You'll notice that the Jersey side of the river is defined by a steep cliff of drab brown rock, about a hundred feet high, studded with vertical cracks. Locals refer to it as the Palisades. During the summer it can be almost unrecognizable, engulfed by a dense forest of trees and bushes that somehow cling to the sheer slopes. Commuter towns like Jersey City and Fort Lee are perched on top of the cliff, and the western end of the George Washington Bridge is built deep into it, an ideal anchor for the world's busiest overwater crossing. If you wanted to, you could walk along the Palisades for about fifty miles, from where it begins in Staten Island and extends along the Hudson to where it juts into upstate New York.

Millions of people look at this cliff every week. Hundreds of thousands of people live on it. Few realize that it is a remnant of those ancient volcanic eruptions that tore apart Pangea and ushered in the Age of Dinosaurs.

The Palisades is what geologists refer to as a sill—an intrusion of magma that pokes its way in between two layers of rock far underground, but then hardens into stone before it can erupt as lava. Sills are part of the internal plumbing system of volcanoes. Before they harden into rock, they are pipes, which transport magma underground. Sometimes they are conduits that bring magma to the surface; other times they are dead-end extensions of the volcanic system, cul-de-sacs that magma can't escape from. The Palisades sill formed at the end of the Triassic, as Pangea was rupturing along what would become the eastern coast of North America, just a few miles from what is now New

York City. It formed from those very magmas that were coursing up from the deep earth as the supercontinent broke into two.

The magma that became the Palisades sill never made it to the surface. It never got to be part of those three-thousand-foot-thick lava sheets flushed out of the Pangean rift, the ones that engulfed ecosystems and belched out the carbon dioxide that would doom much of the planet. About twenty miles to the west the magmas did erupt, however, and the basalt rocks that formed from them can be seen in a low range of hills called the Watchung Mountains in northern New Jersey. Calling them mountains is generous—they're just a few hundred feet high, and they cover a tiny area about forty miles north to south—but they are a beloved oasis of natural beauty within one of the most urbanized parts of the world.

In the middle of the mountains is Livingston, a bedroom community of about thirty thousand people. In 1968 some folks discovered dinosaur footprints a couple of miles north of the town, in an abandoned quarry where red shales, formed in rivers and lakes near the ancient volcanoes, were being mined. There was a blurb in the local newspaper, which caught the eye of a mother, who told her fourteen-year-old-son, Paul Olsen, who was gobsmacked to learn that dinosaurs once lived so close to his home. He rounded up his friend, Tony Lessa, and they hopped on their bicycles and sped to the old quarry. It was no more than an overgrown, rock-strewn hole in the ground, but the discovery had caused a local sensation and several amateur collectors were already there, on the hunt for more tracks. Olsen and Lessa befriended some of the amateurs, who taught them the basics of fossil collecting: how to identify dinosaur footprints, how to remove them from the rock, and how to study them.

The two teenagers became obsessed. They kept coming back to the quarry, and before long they were working late into the night, removing slabs of dinosaur footprints by firelight, even in the dead of winter. They had to go to school during the day, so the night was their only option. For over a year they toiled, outlasting the other rockhounds, who began to trickle away once the excitement of the discovery died down. The boys collected hundreds of tracks left by all kinds of creatures, including meat-eating dinosaurs similar to *Coelophysis* from Ghost Ranch, plant-eating dinosaurs, and some of the scaly and furry creatures that lived alongside. But the more they collected, the more they became dismayed: during their nighttime excavations, they were constantly interrupted by trucks illegally dumping trash, and while they were at school, unscrupulous collectors would often sneak into the quarry and poach footprints the boys hadn't yet been able to remove.

So what's a 1960s teenager to do when his favorite fossil site is being destroyed? Paul Olsen skipped the middlemen and went right to the top. He began writing letters to Richard Nixon, the newly elected president who had yet to disgrace himself. Lots of letters. He begged Nixon to use his presidential powers to get the quarry preserved as a protected park, and even sent a fiberglass cast of a theropod track to the White House. Olsen led a media campaign, too, and was profiled in an article in *Life* magazine. His brazen persistence paid off: in 1970 the company that owned the quarry donated the land to the county, which made it into a dinosaur park called the Riker Hill Fossil Site. The next year, the site was granted official national landmark status and Olsen received a presidential commendation for his work. Little did he know it, but he was also an inch away from a White House visit.

Some of Nixon's image-conscious aides thought a photo op with a young science enthusiast would be great PR for the jowly president, but it was killed at the last minute by Nixon's advisor John Ehrlichman, later one of the key villains of Watergate.

It was a great accomplishment for a kid—collecting a haul of dinosaur tracks, getting his site preserved for posterity, becoming pen pals with the president. But Paul Olsen didn't stop. He went to college to study geology and paleontology, completed a PhD at Yale, and was hired as a professor at Columbia University, across the Hudson from Riker Hill. He became one of the leading academic paleontologists in the world and was elected to the National Academy of Sciences, one of the greatest honors for any American scientist. He also had the burden of being a member of my PhD committee, a far lesser honor, when I did my doctorate, in New York. During that time, he became one of my most trusted mentors, a brilliant sounding board for whatever crazy research ideas I had. For two years, I assisted him as he taught his popular undergraduate course on dinosaurs at Columbia, always oversubscribed by nonmajor students, who were seduced by the eminent scientist with a white Geraldo moustache prancing around with the enthusiasm that comes from several preclass energy drinks. Much of my ebullient, wildly animated lecturing style comes from watching Paul.

Paul Olsen made his career by continuing what he started as a teenager. Much of his work has focused on those events that were occurring around the time dinosaurs were leaving footprints in New Jersey: the breakup of Pangea at the very end of the Triassic, the unimaginable volcanic eruptions, the mass extinction, and the rise of dinosaurs to global dominance as the Triassic transitioned into the subsequent Jurassic Period.

Although he had no idea when he first cycled up to that quarry as a kid, Paul grew up in the best place in the world for studying the Late Triassic and Early Jurassic. His boyhood stomping grounds are within a geological structure called the Newark Basin, a bowl-like depression filled with Triassic and Jurassic rocks. It is one of many such structures—called rift basins, because they formed as Pangea rifted apart—extending for over a thousand miles down the eastern coast of North America. The Bay of Fundy, up north in Canada, laps onto one of these basins. Farther south is the Hartford Basin, which cuts through much of central Connecticut and Massachusetts. Then the Newark Basin, followed by the Gettysburg Basin, site of the famous Civil War battle, the topography of the rocks so instrumental in shaping military strategy that depended on securing bits of high ground. South of Gettysburg are many smaller basins that pepper the backcountry of Virginia and North Carolina, finally culminating in the huge Deep River Basin of the Carolina interior.

These rift basins follow the fracture between east and west Pangea. They are the dividing line, the frontier, the place where the supercontinent tore up. As those east-west tugging forces started to pull Pangea apart, faults formed deep within the crust, cutting through what used to be solid rock. Each bit of tugging would cause an earthquake, which would cause the rocks on either side of the fault to move a little bit relative to each other. Over millions of years the faults reached the surface, and as one side continued to fall, a basin was formed: a depression on the downward side of the fault rimmed by a high mountain range on the upward side. Each of the eastern North American rift basins formed this way, the result of more than 30 million years of pressure, tension, and tremors.

This is exactly what is occurring in eastern Africa today, as Africa is pulling away from the Middle East at the rate of about one centimeter per year. The two landmasses used to be connected about 35 million years ago, but now they are separated by the long and skinny Red Sea, which continues to get wider year by year and will one day turn into an ocean. To the south, on the African mainland, there is a north-south band of basins, each growing wider and deeper with every earthquake that is yanking Africa and Arabia farther apart. Some of the deepest lakes in the world, like the nearly mile-deep Lake Tanganyika, fill some of these basins. Others are crisscrossed by raging rivers, which rush down from the mountains above, irrigating great tropical ecosystems lush with some of Africa's most familiar plants and animals. Sprinkled throughout, poking up in random places, are volcanoes like Mount Kilimanjaro, escape valves for the magma building up underground as the land fractures. Occasionally one of these goes off and buries the basins, and their inhabitants, in lava and ash.

Paul Olsen's Newark Basin, and the many others lining the eastern coast of North America, underwent a similar process of evolution. They were formed gradually by earthquakes, were flushed with rivers that supported diverse ecosystems, eventually became so deep and full of water that the rivers turned into lakes, and then, depending on the quirks of climate, the lakes would dry up, rivers would form again, and the whole process would start over. Cycle after cycle after cycle. Dinosaurs, pseudosuchian cousins of crocodiles, supersalamanders, and early relatives of mammals thrived along the rivers' edges, and blooms of fish choked the lakes. These animals left their fossils—the footprints Paul Olsen started to collect as a teenager, as well as bones—in the thousands of feet of sandstones, mudstones, and other rocks deposited by the rivers and lakes. And then, when Pangea had been stretched to its limit, the crust burst and volcanoes started to erupt, burying the basins and the creatures that lived within them.

The first eruptions didn't occur in the Newark Basin area. They happened in what is now Morocco, which at that time was nudged up against what would become eastern North America, just a few hundred miles or so from modern New York City. Then lava began pouring out in other places where Pangea was splitting: in the Newark Basin, in what is now Brazil, in those same lake environments where we found the supersalamander graveyard in Portugal—all along that zipper line, which, many millions of years later, would transform into the Atlantic Ocean. The lava came in four waves, each scorching the once verdant rift basins, each spreading toxic fumes all over the planet, each making a bad situation worse and worse. In only about half a million years—a blink of an eye in geological terms— the eruptions stopped, but they transformed the Earth forever.

The dinosaurs, pseudosuchian crocodile-line archosaurs, big amphibians, and early mammal relatives living in the rift basins were blissfully unaware of what was about to happen. Things went sour quickly.

The initial eruptions in Morocco released clouds of carbon dioxide, a powerful greenhouse gas, which rapidly warmed the planet. It got so hot that strange ice formations buried within the seafloor, called clathrates, melted in unison all throughout the world's oceans. Clathrates are unlike the solid blocks of ice we're used to, the ones we put in our drinks or carve into fancy sculptures at parties. They are a more porous substance, a latticework of frozen water molecules that can trap other substances inside it. One of those substances is methane, a gas that seeps up constantly from the deep Earth and infiltrates the oceans but is caged in the clathrates before it can leak into the atmosphere. Methane is nasty: it's an even more powerful greenhouse gas than carbon dioxide, packing an earth-warming punch over thirty-five times as great. So when that first torrent of volcanic carbon dioxide increased global temperatures and melted the clathrates, all of that once-trapped methane was suddenly released. This initiated a runaway train of global warming. The amount of greenhouse gas in the atmosphere approximately tripled within a few tens of thousands of years, and temperatures increased by 3 or 4 degrees Celsius.

Ecosystems on land and in the oceans couldn't cope with such rapid change. The much hotter temperatures made it impossible for many plants to grow, and indeed upwards of 95 percent of them went extinct. Animals that fed on the plants found themselves without food, and many reptiles, amphibians, and early mammal relatives died out, like dominoes falling up the food chain. Chemical chain reactions made the ocean more acidic, decimating the shelly organisms and collapsing food webs. Climate became dangerously variable, with episodes of intense heat followed by cooler periods. This enhanced the temperature differences between northern and southern Pangea, causing the megamonsoons to become more severe, the coastal regions to become even wetter, and the continental interiors to grow much drier. Pangea had never been a particularly hospitable place, but those early dinosaurs that already were constrained by the monsoons, the deserts, and their pseudosuchian rivals were now in even worse shape.

So how did these dinosaurs, still at such a relatively young stage in their evolution, deal with a world that was changing so quickly? The clues are in the footprints that Paul Olsen has been studying now for nearly fifty years. The quarry that Paul explored in New Jersey is one of more than seventy places where dinosaur footprints have been found along the eastern seaboard of the US and Canada. These sites are positioned one on top of the other, in geological sequence, stretching over 30 million years, from around the time the first dinosaurs were originating in what is now South America (but still absent in modern-day North America), through the Late Triassic, across the volcanic extinction, and into the ensuing Jurassic Period. Generations of dinosaurs and other animals left their traces in those cyclical beds of sandstones and mudstones deposited in the rift basins, and by studying them in succession, you can see how these creatures were evolving.

The rocks tell a remarkable story. During the Late Triassic, beginning about 225 million years ago, when the rift basins were just beginning to form, dinosaurs started to leave their marks in the form of rare footprints. There are three-toed tracks called *Grallator*, ranging from about two to six inches (five to fifteen centimeters) long, made by small, fast-running, meat-eating dinosaurs that stood on two legs like the Ghost Ranch *Coelophysis*. There's a second type of track called *Atreipus*, which are about the same size as *Grallator* but include small handprints next to the three-toed footprints, a sign that the trackmaker was walking on all fours. They were probably made by primitive ornithischian dinosaurs—the oldest cousins of *Triceratops* and the duck-billed dinosaurs—or perhaps by close dinosauromorph cousins to dinosaurs. These dinosaur tracks are vastly outnumbered by the prints left by pseudosuchians, large amphibians, proto-mammals, and small lizards. Dinosaurs were there, but they still remained role players in the rift basin ecosystems, right up until the end of the Triassic.

But then the volcanoes kicked into gear. Suddenly the diversity of non-dinosaur tracks drops dramatically in those first Jurassic rocks above the lava flows. Many non-dinosaur tracks abruptly disappear, including some of the most conspicuous prints left by crocodile-cousin pseudosuchians, which had previously been more abundant and diverse than the dinosaurs. Whereas dinosaurs made up only about 20 percent of all tracks before the volcanoes, right afterward half of all footprints belong to dinosaurs. A variety of totally new dinosaur tracks enter the record: a handprint-footprint duo called *Anomoepus* probably made by an ornithischian, a large four-toed print called *Otozoum* made by the very first long-necked proto-sauropods to live in the rift valleys, and a three-toed track called *Eubrontes* that belonged to another type of swift predator. These *Eubrontes* tracks are a little over a foot long (about thirty-five centimeters), a big size increase over the *Grallator* prints left by similar but much smaller carnivores during the prevolcano days of the Triassic.

It's probably not what you were expecting. After some of the largest volcanic eruptions in Earth history desecrated ecosystems, dinosaurs became *more* diverse, *more* abundant, and larger. Completely new dinosaur species were evolving and spreading into new environments, while other groups of animals went extinct. As the world was going to hell, dinosaurs were thriving, somehow taking advantage of the chaos around them.

When the volcanoes ran out of lava and their six-hundred-thousand-year reign of terror was over, the world was a very different place than it had been in the Late Triassic. It was much warmer, storms were more intense, and wildfires ignited with ease; new types of ferns and ginkgos had replaced the once abundant broadleaf conifers; and many of the most charismatic Triassic animals were gone. The piglike mammal-relative dicynodonts and the beaked plant-munching rhynchosaurs were both extinct; the supersalamander amphibians, almost completely knocked out. What about the pseudosuchians, those crocodile-line archosaurs that were overshadowing, outmuscling, and seemingly outcompeting the dinosaurs during the final 30 million years of the Triassic? Nearly every species bit the dust. The long-snouted phytosaurs, the tanklike aetosaurs, the apex-predator rauisuchians, and the weird *Effigia*-like critters that resembled dinosaurs—none of them were ever to be heard from again. The only pseudosuchians that made it through the great Pangean breakup were a few types of primitive crocodiles, a handful of battle-worn stragglers that would eventually evolve into the modern alligators and crocodiles but would never enjoy the same success they had in the Late Triassic, when they seemed primed to take over the world.

Somehow dinosaurs were the victors. They endured the Pangean split, the volcanism, and the wild climate swings and fires that vanquished their rivals. I wish I had a good answer for why. It's a mystery that quite literally has kept me up at night. Was there something special about dinosaurs that gave them an edge over the pseudosuchians and other animals that went extinct? Did they grow faster, reproduce quicker, have a higher metabolism, or move more efficiently? Did they have better ways of breathing, hiding, or insulating their bodies during extreme heat and cold snaps? Maybe, but the fact that so many dinosaurs and pseudosuchians looked and behaved so similarly makes such ideas tenuous at best. Maybe dinosaurs were just lucky. Perhaps the normal rules of evolution are ripped up when such a sudden, devastating, global catastrophe happens. It could be that the dinosaurs simply were the ones that walked away from the plane crash unscathed, saved by good fortune, when so many others died.

Whatever the answer, it's a riddle waiting for the next generation of paleontologists to figure out.

THE JURASSIC PERIOD marks the beginning of the Age of Dinosaurs proper. Yes, the first true dinosaurs entered the scene at least 30 million years before the Jurassic began. But as we've seen, these earlier Triassic dinosaurs had not even a remote claim to being dominant. Then Pangea began to split, and the dinosaurs emerged from the ashes and found themselves with a new, much emptier world, which they proceeded to conquer. Over the first few tens of millions of years of the Jurassic, dinosaurs diversified into a dizzying array of new species. Entirely new subgroups originated, some of which would persist for another 130-plus million years. They got larger and spread around the globe, colonizing humid areas, deserts, and everything in between. By the middle part of the Jurassic, the major types of dinosaurs could be found all over the world. That quintessential image, so often repeated in museum exhibits and kids' books, was real life: dinosaurs thundering across the land, at the top of the food chain, ferocious meat-eaters comingling with long-necked giants and armored and plated plant-eaters, the little mammals and lizards and frogs and other non-dinosaurs cowering in fear.

Here are some of the familiar dinosaurs that start to show up after the Pangean rift volcanoes ushered in the Jurassic. There were meat-eating theropods like *Dilophosaurus*, with a weird double-mohawk crest on its skull; at around twenty feet long, it was much larger than the mule-size *Coelophysis* and most other Triassic carnivores. Plant-eating ornithischians covered in armor plates, like *Scelidosaurus* and *Scutellosaurus*, would soon after give rise to the familiar tanklike ankylosaurs and back-plated stegosaurs. Small, fast-moving, probably omnivorous ornithischians like *Heterodontosaurus* and *Lesothosaurus*, were early members of that lineage that would eventually produce the horned and duck-billed dinosaurs. Other familiar dinosaurs that had been around in the Triassic but restricted to only a few environments, like the long-necked proto-sauropods and the most primitive ornithischians, finally began to migrate around the planet.

Nothing in this inventory of growing diversity encapsulates the newfound dominance of dinosaurs quite like the sauropods. They are those unmistakable long-necked, column-limbed, potbellied, plant-devouring, small-brained behemoths. Some of the most famous dinosaurs of all are sauropods: *Brontosaurus*, *Brachiosaurus*, *Diplodocus*. They show up in almost all museum exhibits and are stars of *Jurassic Park*; Fred Flintstone used one to mine slate, and a green cartoon sauropod has been the logo of Sinclair Oil for decades. Along with *T. rex*, they are *the* iconic dinosaurs.

Sauropods evolved from an ancestral stock, what I've been calling the proto-sauropods, in the latest Triassic. These proto-species were the dog-to-giraffe-size plant-eaters with fairly long necks that were among that first wave of dinosaurs to appear in Ischigualasto about 230 million years ago. They then became

The skull of *Plateosaurus*, one of the proto-sauropods, the ancestral stock that gave rise to the sauropods.

the main herbivores in the humid parts of Triassic Pangea but were kept from achieving their full potential by their inability to settle the deserts. That changed in the early part of the Jurassic, when sauropods were able to break free of their environmental restrictions and move about the globe, evolving their characteristic noodle-necked bodies and growing to monstrous sizes in the process.

Fossils of some of the first truly gigantic sauropods—ones that weighed over ten tons, were over fifty feet long, and had necks that could stretch several stories into the sky—have started turning up in Scotland over the past few decades, on a beautiful island off the west coast called the Isle of Skye. The clues have been meager—a stocky limb bone here, a tooth or a tail vertebra there—but they hint at an animal of enormous size living about 170 million years ago, far enough into the Jurassic that the Pangean split and volcanic apocalypse were distant memories but still during that time when dinosaurs were putting the final flourishes on their rise to dominance.

The sauropod fossils from Skye piqued my interest when I moved to Scotland in 2013 to take up my new position at the University of Edinburgh, fresh out of my PhD in New York and bounding with the excitement of starting my own research lab. During my first few weeks on the job, I began to hang out with two scientists in my department: Mark Wilkinson, a hardened field geologist whose ponytail and scruffy beard give him the look of a hippie, and Tom Challands, a redheaded packhorse of man who also had a doctorate in paleontology, albeit on microscopic fossils from over 400 million years ago. Tom had recently finished a stint in the real world, putting his geological skills to use working for an energy company on the search for oil. For part of that time, he lived in a custom-made camper van, fitted with a bed and small kitchen, which he would park near whatever sites he was surveying. His new bride put the kibosh on that lifestyle after their wedding, but the van still came in handy for fieldwork travel, and Tom would often spend his weekends driving along the misty coasts of Scotland looking for whatever fossils he could find. Both Tom and Mark had done some geological work on Skye and knew the terrain well, so we made a pact to hunt for better fossils of the mysterious giant sauropods.

The more we read about Skye, the more one name kept popping up: Dugald Ross. It was a name I wasn't familiar with. He wasn't a paleontologist or a geologist or a scientist of any kind. Yet he had discovered and described many of the dinosaur fossils found on Skye. Dugald was a local boy who grew up in the tiny hamlet of Ellishadder on the far northeastern arm of the island, a rugged landscape of craggy peaks, green hills, peat-colored streams, and windswept shores that looks like something out of a fantasy novel—very Tolkienesque. He was raised in a household that spoke Gaelic, the native language of the Scottish Highlands, which is spoken by only about fifty thousand people today but which still has a presence on the road signs and in the schools on remote islands like Skye. When Dugald was fifteen years old, he found a cache of arrow points and Bronze Age artifacts near his family's home, and this sparked an obsession with the history of his native island that continued into adulthood, as he carved out a career as a builder and crofter (a Scottish Highlands term for a small-scale farmer and sheepherder).

I got in touch with Dugald and told him about our dreams of finding huge dinosaurs on his island. It was one of the most fortunate e-mails I've ever sent, because it struck up a friendship and a remarkable scientific collaboration. Dugald—or Dugie, as he prefers to be called—invited us to visit him when we came up to his island a few months later. He instructed us to drive up the main two-lane road that snakes along the coast of northeastern Skye and meet him at a long ranch-style building, made up of a collage of different-size gray stones and a black tile roof, with antique farming instruments strewn across the lawn outside. There was a sign out front that said TAIGH-TASGAIDH—the Gaelic word for museum. Dugie emerged from his big red work van with a set of oversize skeleton keys, made his introductions, and proudly led us inside. In his soft-spoken lyrical accent—a

The enchanting landscape of the Isle of Skye, Scotland.

charming combination of Sean Connery–style Scots and an Irish brogue—he explained how he had taken the ruins of a one-room schoolhouse and built the structure we were standing in, the Staffin Museum. He founded the museum when he was nineteen. Today, this single room—without a café, a big gift shop, or other expensive trappings of big-city museums, or even electricity—contains many of the dinosaurs he's found on Skye, along with artifacts that trace the history of the island's human inhabitants. It's a surreal experience: big dinosaur bones and footprints displayed next to old mill wheels, iron rods for picking turnips, and antique mole traps once used by Highland farmers.

For the rest of that week, Dugie led us to many of his favorite hunting spots. We found a lot of Jurassic fossils—the jaw of a

Dugie Ross removing a dinosaur bone from a boulder on the Isle of Skye.

The dinosaur dance floor of sauropod tracks that I discovered with Tom Challands on the Isle of Skye.

dog-size crocodile, the teeth and backbones of reptiles called ichthyosaurs, which resembled dolphins and lived in the oceans when dinosaurs started to dominate the land—but no giant sauropods. Over the next few years, we kept coming back.

Finally, in the spring of 2015, we found what we set out for, although we didn't even realize it at first. We spent most of the day on our hands and knees, looking for tiny fish teeth and scales embedded in a platform of Jurassic rocks that stretched into the icy waters of the North Atlantic, right below the ruins of a fourteenth-century castle. This was Tom's idea: he now studies fossil fish, and in exchange for his help finding dinosaurs, I promised to assist him in collecting fishy bits. We had been squinting at the rocks for hours, bundled up in three layers of waterproof clothing but still freezing. The tide was coming in, the late afternoon light was going down, and dinner was beckoning. So Tom and I packed up our gear and our bags of fish teeth and started to stroll back to his tricked-out van parked on the other side of the beach. That's when something caught our eyes. It was a malformed depression in the rock, about the size of a car tire. We had missed it earlier because our eyes were focused on the much smaller fish bones, our search image totally unsuitable for noticing something so big.

As we continued to walk, we started to notice many other similar depressions, now visible in the low-angle afternoon light. They were all about the same size, and the closer we looked, the more we saw that they stretched in every direction around us. They seemed to show a pattern. Individual holes were lined up in two long rows, in something of a zigzag arrangement: left-right, left-right, left-right. Ribbons of them were crisscrossing much of the rock platform that we had been working on all day.

Tom and I looked at each other. It was the kind of knowing glance between brothers, a nonverbal connection based on years of shared experience. We had seen these types of things before, not in Scotland, but in places like Spain and western North America. We knew what they were.

The holes in front of us were fossilized tracks, huge ones. Dinosaur tracks, no doubt. As we looked closer, we could see that there were both handprints and footprints, and some of them had finger and toe marks. They had the telltale shape of tracks left by sauropods. We had found a 170-million-year-old dinosaur dance floor, records left by colossal sauropods that were about fifty feet long and weighed as much as three elephants.

The tracks were made in an ancient lagoon, an environment not commonly associated with sauropods. We usually envision these monstrous dinosaurs stampeding across the land, causing a small earthquake with each step. And they did. But by the middle part of the Jurassic, the sauropods had become so diverse that they started branching out into other ecosystems, always searching for the vast quantities of leafy food needed to fuel their giant bodies. Our trackway site in Skye has at least three different layers of footprints, made by different generations of sauropods wading through a salty lagoon, living with smaller plant-eating dinosaurs, the occasional pickup-truck-size carnivore, and many types of crocodiles, lizards, and swimming mammals with flat tails like beavers. Scotland was much warmer back then, a land of swamps and sandy beaches and rolling rivers on an island in the middle of the growing Atlantic Ocean, perched between North American and European landmasses that moved farther and farther apart as Pangea continued to split. Thoroughly ruling this land were the sauropods and other dinosaurs, which had now—finally—become a global phenomenon.

THERE'S REALLY NO better way to say it: the sauropods that made their marks in that ancient Scottish lagoon were awesome creatures. Awesome in the literal sense of the word—impressive, daunting, inspiring awe. If I was handed a blank sheet of paper and a pen and told to create a mythical beast, my imagination could never match what evolution created in sauropods. But they were real: they were born, they grew, they moved and ate and breathed, they hid from predators, they slept, they left footprints, they died. And there's absolutely nothing like sauropods around today—no animals with a similar long-necked and swollen-gut body type, no creatures on land that even remotely approach them in size.

Sauropods are so mind-twistingly big that, when their first fossil bones were discovered in the 1820s, scientists found themselves in a bind. Some of the first dinosaurs were being found around the same time, like the meat-eating *Megalosaurus* and the beaked herbivore *Iguanodon*. These were big animals, no doubt, but nowhere near the size of the creatures that left the gigantic sauropod bones. So scientists didn't make the connection with dinosaurs. Instead, they considered the sauropod bones to belong to the one type of thing they knew could get so huge: whales. It was a few decades before that mistake was corrected. Amazingly, later discoveries would show that many sauropods got even bigger than most whales. They were the largest animals that ever walked the land, and they push the limit for what evolution can achieve.

This raises a question that has fascinated paleontologists for over a century: how did sauropods become so large?

It's one of the great puzzles of paleontology. But before trying to solve it, we first need to come to grips with a more fundamental issue: how big did sauropods get? How long were they, how high could they stretch their necks, and most important, how much did they weigh? These turn out to be difficult questions to answer, particularly when it comes to weight, because you can't just stick a dinosaur on a scale and weigh it. A trade secret among paleontologists is that many of the fantastical numbers you see in books and museum exhibits—*Brontosaurus* weighed a hundred tons and was bigger than a plane!—are pretty much just made up. Educated guesses or, in some cases, barely that. Recently, however, paleontologists have come up with two different approaches to more accurately predict the weight of a dinosaur based on its fossil bones.

The first is really quite simple and relies on basic physics: heavier animals require stronger limb bones to support their weight. This logical principle is reflected in how animals are built. Scientists have measured the limb bones of many living animals, and it turns out that the thickness of the main bone in each limb that supports the animal—the femur (thighbone) for those that walk on two legs only or the femur plus the humerus (upper arm bone) for those that stand on all fours—is strongly statistically correlated with the weight of the animal. In other words, there is a basic equation that works for almost all living animals: if you can measure limb-bone thickness, you can then calculate body weight with a small but recognized margin of error—simple algebra you can do with a basic calculator.

The second method is more intensive but a lot more interesting. Scientists are starting to build three-dimensional digital models of dinosaur skeletons, add on the skin and muscles and internal organs in animation software, and use computer programs to calculate body weight. It's a method pioneered by a number of young British paleontologists—Karl Bates, Charlotte Brassey, Peter Falkingham, and Susie Maidment—and their network of collaborators, who include everyone from biologists specializing in living animals to computer scientists and programmers.

A few years ago, when I was finishing my PhD, Karl and Peter invited me to take part in a study of sauropod body size and proportions using digital models. It was an ambitious goal: make detailed computer animations of all sauropods with complete enough skeletons and figure out how big these animals were and how their bodies changed as they grew into truly titanic sizes. I was invited for purely practical reasons: some of the best sauropod skeletons in the world are on display at the American Museum of Natural History in New York City, where I was based at the time, and they needed data for one of them in particular, a Late Jurassic species called *Barosaurus*. They instructed me how to gather the information to build the model, and I was surprised that all it required was a normal digital camera, a tripod, and a scale bar. I took about a hundred photos of the *Barosaurus* skeletal mount from all possible angles, keeping my camera steady on the tripod and making sure to include a ruler in most of the images. Then Karl and Peter input the images into a computer program that matches equivalent points on the

Brontosaurus at the American Museum of Natural History in New York, with a human skeleton for scale. *American Museum of Natural History Library; see copyright page for full AMNH library credit information.*

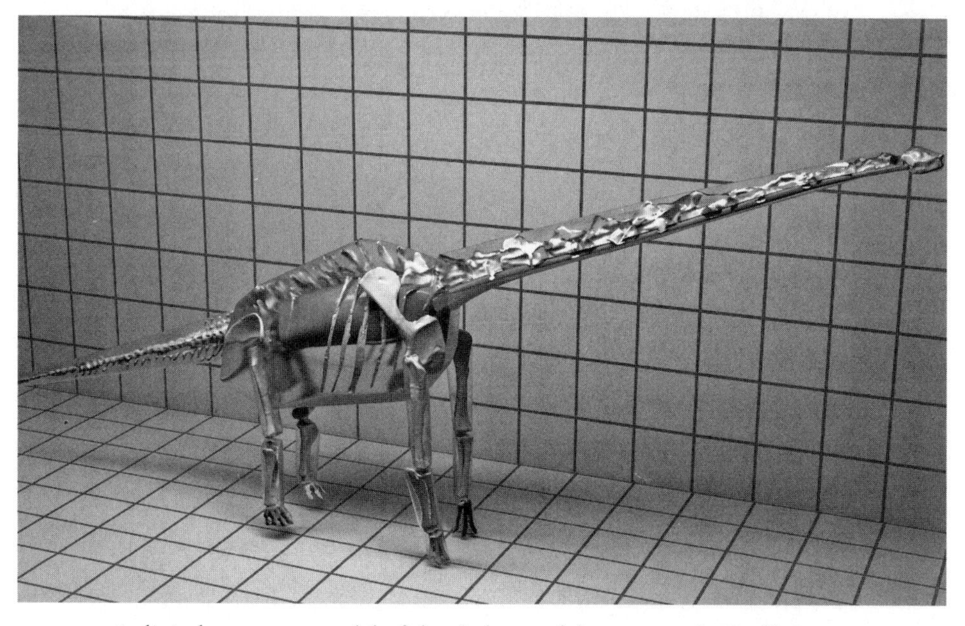

A digital computer model of the skeleton of the sauropod *Giraffatitan*, which helps scientists calculate the weight of the animal. *Courtesy of Peter Falkingham and Karl Bates.*

photographs, works out the distances between them based on the scale, and does this continuously until a three-dimensional model is built from the original 2-D images.

The technique is called photogrammetry, and it's revolutionizing how we study dinosaurs. The super-accurate models it creates can be measured in precise detail. Or they can be loaded into animation software and made to run and jump, in order to determine what kinds of motions and behaviors dinosaurs were capable of. They can even be used to animate movies or television documentaries, ensuring that the most realistic dinosaurs appear on screen. These models are bringing dinosaurs to life.

Our computer modeling study and more traditional studies based on limb-bone measurements come to the same conclusion: sauropod dinosaurs were really, really big. The primitive proto-sauropods like *Plateosaurus* began to experiment with relatively large sizes in the Triassic, as some of them got up to about two or three tons in weight. That's roughly equivalent to a giraffe or two. But after Pangea started to split, the volcanoes erupted, and the Triassic turned into the Jurassic, the true sauropods got much larger. The ones that left tracks in the Scottish lagoon weighed about ten to twenty tons, and later in the Jurassic, famous beasties like *Brontosaurus* and *Brachiosaurus* expanded to more than thirty tons. But that was nothing compared to some supersize Cretaceous species like *Dreadnoughtus, Patagotitan, Argentinosaurus*—members of an aptly named subgroup called the titanosaurs—which weighed in excess of fifty tons, more than a Boeing 737.

The biggest and heaviest land animals today are elephants. Their sizes vary, depending on where they live and which species they belong to, but most weigh about five or six tons.

Apparently the largest one ever recorded was around eleven tons. They have nothing on sauropods. Which circles back to the money question: how were these dinosaurs able to attain sizes so completely out of scale with anything else evolution has ever produced?

The first thing to consider is what animals require to become really big. Perhaps most obvious, they need to eat a lot of food. Based on their sizes and the nutritional quality of the most common Jurassic foodstuffs, it's estimated that a big sauropod like *Brontosaurus* probably needed to eat around a hundred pounds of leaves, stems, and twigs every day, maybe more. So they needed a way to gather and digest such vast quantities of grub. Secondly, they need to grow fast. Growing bit by bit, year by year is all well and good, but if it takes you over a century to get big, that's many opportunities for a predator to eat you, or a tree to fall on you during a storm, or a disease to take you out long before you grow into your full-size adult body. Third, they must be able to breathe very efficiently, so they can take in enough oxygen to power all of the metabolic reactions in their immense bodies. Fourth, they need to be constructed in a way that their skeleton is strong and sturdy, but also not so bulky that it can't move. Finally, they need to shed excess body heat, because in hot weather it is very easy for a big creature to overheat and die.

Sauropods must have been able to do all of these things. But how? Many scientists who started to ponder this riddle decades ago went for the easiest answer: maybe there was something different in the physical environment back in the Triassic, Jurassic, and Cretaceous. Perhaps gravity was weaker, so heftier animals could move and grow more easily back then. Or maybe there was more oxygen in the atmosphere, so the hulking sauropods could breathe, and therefore grow and metabolize, more efficiently. These speculations might sound convincing, but on closer scrutiny they don't check out. There is no evidence gravity was substantially different during the Age of Dinosaurs, and oxygen levels back then were about the same as today, or maybe even slightly lower.

That leaves only one plausible explanation: there was something intrinsic about sauropods that allowed them to break the shackles that constrained all other land animals—mammals, reptiles, amphibians, even other dinosaurs—to much smaller sizes. The key seems to be their unique body plan, which is a mixture of features that evolved piecemeal during the Triassic and earliest Jurassic, culminating in an animal perfectly adapted for thriving at large size.

It all starts with the neck. The long, spindly, slinky-shaped neck is probably the single most distinctive feature of sauropods. A longer-than-normal neck started to evolve in the very oldest Triassic proto-sauropods, and it got proportionally longer over time, as sauropods both added more vertebrae—the individual bones in the neck—and stretched each individual vertebra ever further. Like Iron Man's armor, the long necks conferred a kind of superpower: they allowed sauropods to reach higher in the trees than other plant-eating animals, giving them access to a whole new source of food. They could also park themselves in one area for several hours and extend their necks up and down and all around like a cherry picker, gobbling up plants while expending very little energy. That meant they were able to eat more food, and thus take in energy more efficiently, than their competitors. That's adaptive advantage number one: their necks permitted them to eat the huge meals necessary to put on excessive weight.

Then there's the way that they grew. Recall that the dinosauromorph ancestors of dinosaurs developed higher metabolisms, faster growth rates, and a more active lifestyle than many of the amphibians and reptiles that were also diversifying in the earliest Triassic. They weren't lethargic, and it didn't take them aeons to grow into adults like an iguana or a crocodile. This was also true of all of their dinosaur descendants. Studies of bone growth indicate that most sauropods matured from guinea-pig-size hatchlings to airplane-size adults in only about thirty or forty years, an incredibly short period of time for such a remarkable metamorphosis. That's advantage two: sauropods obtained the fast growth essential to reach large size from their distant, cat-size ancestors.

Sauropods also retained something else from their Triassic ancestors: a highly efficient lung. The lungs of sauropods were very similar to those of birds and very different from ours. While mammals have a simple lung that breathes in oxygen and exhales carbon dioxide in a cycle, birds have what is called a unidirectional lung: air flows across it in one direction only, and oxygen is extracted during both inhalation and exhalation. The bird-style lung is extra efficient, sucking up oxygen with each breath in and each exhalation. It's an astounding feature of biological engineering, made possible by a series of balloonlike air sacs connected to the lung, which store some of the oxygen-rich air taken in during inhalation, so that it can be passed across the lung during exhalation. Don't worry if it sounds confusing: it is such a strange lung that it took biologists many decades to figure out how it works.

We know that sauropods had such a birdlike lung because many bones of the chest cavity have big openings, called pneumatic fenestrae, where the air sacs extended deep inside. They are exactly the same structures in modern birds, and they can only be made by air sacs. So that's adaptation three: sauropods had ultra-efficient lungs that could take in enough oxygen to stoke their metabolism at huge size. Theropod dinosaurs had the same bird-style lungs, which could have been one factor that allowed tyrannosaurs and other giant hunters to get so large, but the ornithischian dinosaurs did not. This is why duck-billed dinosaurs, stegosaurs, horned species, and armored dinosaurs were never able to grow as huge as sauropods.

It turns out that air sacs also have another function. Aside from storing air in the breathing cycle, they also lighten the skeleton when they invade bone. In effect, they hollow out the bone, so that it still has a strong outer shell but is much more lightweight, the way an air-filled basketball is lighter than a rock of similar size. Want to know how sauropods could hold up their long necks without toppling over like an unbalanced seesaw? It's because all of the vertebrae were so engulfed by air sacs that they were little more than honeycombs, featherweight but still strong. And that's advantage four: the air sacs allowed sauropods to have a skeleton that was both sturdy and light enough to move around. Without air sacs, mammals, lizards, and ornithischian dinosaurs had no such luck.

And what about the fifth special adaptation, being able to expel excess body heat? The lungs and air sacs helped with this too. There were so many air sacs, and they extended throughout so much of the body, snaking their way into bones and between internal organs, that they provided a large surface area for dissipating heat. Each hot breath would be cooled by this central air conditioning system.

Putting it all together, that's how you can build a supergiant dinosaur. If sauropods had lacked any one of these features—the long neck, the fast growth rates, the efficient lung, the system of skeleton-lightening and body-cooling air sacs—then they probably would not have been capable of becoming such behemoths. It wouldn't have been biologically possible. But evolution assembled all of the pieces, put them together in the right order, and when the kit was finally assembled in the post-volcanic world of the Jurassic, sauropods suddenly found themselves able to do something no other animals, before or since, have been able to do. They became biblically huge and swept around the world; they became dominant in the most magnificent way—and they would remain so for another hundred million years.

DINOSAURS *and* DRIFTING CONTINENTS

Stegosaurus

NESTLED WITHIN THE LEAFY STREETS of New Haven, Connecticut, on the northern fringes of the Yale University campus, there is a shrine. The Great Hall of Dinosaurs at Yale's Peabody Museum may not bill itself as a place of spiritual pilgrimage, but that's sure what it feels like to me. I get a shiver, as when I walked into Catholic mass as a child. It's not a normal shrine—no statues of deities, flickering candles, or the hint of incense. It's also not particularly magnificent, at least from the outside, tucked away inside a fairly nondescript brick building that blends in with the rest of the university's lecture halls. But it houses relics that, to me, are as sacred as those you'll find in most any religious shrine: dinosaurs. To me, there is nowhere better, anywhere on the planet, to go and immerse yourself in the wonder of the prehistoric world.

The Great Hall was originally built in the 1920s to house Yale's incomparable dinosaur collection, assembled over many decades by roughnecks who fanned across the American West and, for the right fee, sent fossil treasures eastward to be studied by the Ivy League elite. Coming up on its centennial, the gallery retains all of its original charm. This isn't some New Age exhibit space with flashing computer screens and dinosaur holograms and a roaring soundtrack in the background. It's a temple of science, where skeletons of some of the most iconic dinosaurs stand in solemn vigil, lights down low, in the sort of silence you really do expect in a church.

Covering the entire east wall is a mural that stretches more than a hundred feet long and sixteen feet high. Taking four and a half years to complete, it was painted by a man named Rudolph Zallinger, who was born in Siberia, moved to the United States, and took up illustration professionally during the Great Depression. If he were around today, Zallinger would probably be working for an animation studio as a storyboard artist. He was a master at setting scenes and incorporating diverse sets of characters, telling grandiose stories with the stroke of his brush. His most famous work is undoubtedly *The March of Progress*—that often satirized timeline of human evolution in which a knuckle-walking ape gradually morphs into a spear-carrying man. More people have probably come to understand, or misunderstand, the theory of evolution through that one image than through all of the textbooks, school lectures, and museum exhibits the world over.

But before he was painting humans, Zallinger was obsessed with dinosaurs. His mural inside the Great Hall—called *The Age of Reptiles*—is the crowning achievement of that stage of his career. It's been on US postage stamps, was featured in a *Life* magazine series, and is either reproduced or plagiarized on all sorts of dinosaur paraphernalia. It's the Mona Lisa of paleontology, surely the single most talked-about piece of dinosaur artwork that has ever been created. But really, it's more akin to the Bayeux Tapestry, because it tells an epic tale of conquest. It's the saga of how fishy creatures first emerged onto land, colonized a new environment, and diversified into reptiles and amphibians; then, of how these reptiles split off into the mammal and lizard lines, the proto-mammals having their day and the lizards following, eventually producing the dinosaurs.

As the mural nears its end, some sixty feet and 240 million years from where it started, after a long journey through alien landscapes of primeval scaly beasts, the painting finally

The theropod *Deinonychus* stands guard over the Zallinger mural at the Peabody Museum, Yale University.

becomes engulfed in dinosaurs. It kind of sneaks up on you, as the transition from the lizards and proto-mammals to the dinosaurs unfolds incrementally across the canvas. Now it's dinosaurs everywhere, of all shapes and sizes, some enormous and others blending into the background. Suddenly, the mural has taken on the feel of something quite different—of a Soviet propaganda poster with Stalin gesticulating before a crowd of peasants, or one of those hilariously self-aggrandizing frescoes in Saddam's palaces. One glance at the dinosaurs and I feel the power. Strength, control, dominance. The dinosaurs were in command, and this was their world.

This part of Zallinger's mural beautifully encapsulates what it was like when dinosaurs had ascended to the peak of their evolutionary success. A monstrous *Brontosaurus* lounges in a swamp in the foreground, munching away on the ferns and evergreen trees surrounding the water. Off to the side, a bus-size *Allosaurus* rips into a bloodied carcass with its teeth and claws, its massive feet stomping on its prey for a little extra insult. Keeping a safe distance is a peaceful grazing *Stegosaurus*, which displays its full arsenal of bony plates and spikes just in case the carnivore has other ideas. Far in the background, where the swamp disappears into a wall of snowcapped mountains, another sauropod uses its long neck to vacuum shrubs off of the ground. Meanwhile, two pterosaurs—those flying reptiles closely related to dinosaurs, often called pterodactyls—chase each other overhead, dipping and diving through the tranquil blue sky.

Odds are, this is the type of image that many of us think of when we think of dinosaurs. These are dinosaurs at their pinnacle.

ZALLINGER'S MURAL IS not fiction. Like any good art, it takes a few liberties here and there, but it is largely rooted in fact. It's based on those very same dinosaurs that stand in front of it in the Great Hall: familiar names like *Brontosaurus*, *Stegosaurus*, and *Allosaurus*. These dinosaurs lived during the Late Jurassic Period, about 150 million years ago. By this time, dinosaurs had already become the dominant force on land. Their victory over the pseudosuchians was 50 million years in the rearview mirror, and it had been a good 20 million years since some of the first giant long-necked species were splashing through the lagoons of Scotland. Nothing was holding back the dinosaurs anymore.

We know a lot about the dinosaurs of the Late Jurassic. That's because there are abundant fossils from this time, in many parts of the world. It's just one of those quirks of geology: some time periods are better represented in the fossil record than others. It's usually because more rocks were being formed during that time, or rocks of that age have better survived the rigors of erosion, flooding, volcanic eruptions, and all of the other forces that conspire to make fossils difficult to find. When it comes to the Late Jurassic, we enjoy two lucky breaks. First, there were hugely diverse communities of dinosaurs living alongside rivers, lakes, and seas all around the world—the perfect places to bury fossils in sediments that later turned to rock. Second, these rocks are today exposed in places convenient for paleontologists—in sparsely populated and dry regions of the United States, China, Portugal, and Tanzania, where annoyances like buildings, highways, forests, lakes, rivers, and oceans don't cover up the fossil booty.

The most famous Late Jurassic dinosaurs—those in Zallinger's mural—come from a thick rock deposit that pokes out all across the western United States. Its technical term is the Morrison Formation, named for a small town in Colorado where there are some beautiful exposures of its colorful mudstones and beige-tinged sandstones. The Morrison Formation is a monster: it can be found in thirteen states today, covering nearly four hundred thousand square miles (a million square kilometers) of the American scrublands. It is easily sculpted into low hills and undulating badlands, the sort of classic backdrop you see in Western films. It's also the source rock for some of the country's most important uranium ore deposits. And, yes, it's a hotbed of dinosaurs, ones whose uranium-infused bones make Geiger counters sing.

Paul Sereno in Wyoming.

Excavating sauropod bones in the Morrison Formation near Shell, Wyoming. At the center back is Sara Burch, who later became an expert on *T. rex* arms (see Chapter 6).

I worked in the Morrison Formation for two summers as an undergraduate. It's where I cut my teeth excavating dinosaur skeletons. I was apprenticing in the lab of the University of Chicago's Paul Sereno, whom we last met leading the expeditions to Argentina that turned up some of the world's very oldest dinosaurs, the Triassic-age *Herrerasaurus*, *Eoraptor*, and *Eodromaeus*. But Paul seemed to study everything and do fieldwork everywhere: he had also found bizarre fish-eating and long-necked dinosaurs in Africa, he'd explored China and Australia, and he'd even described important fossils of crocodiles, mammals, and birds.

In addition, like any academic paleontologist, Paul also had to spend time in the classroom. Each year he taught a popular undergraduate class called Dinosaur Science, which combined theory with practice. Because you can't find dinosaurs anywhere near Chicago, the class would take a ten-day field trip each summer to Wyoming, where the students had the once-in-a-lifetime opportunity to dig dinosaurs with a celebrity scientist. Although at the time I had little prior experience, I was brought on as a teaching assistant, Paul's right-hand man as we herded the students—a diverse lot, from premeds to philosophy majors—across the high desert.

Paul's field sites were located near the tiny town of Shell, secluded between the Bighorn Mountains to the east, and Yellowstone National Park a hundred miles to the west. Only eighty-three people were counted during the last census. When we were there in 2005 and 2006, the road signs boasted of merely fifty residents. But that's a good thing for paleontologists. The fewer people in the way of the fossils, the better. And although Shell is a forgettable dot on the map, it can rightly stake its claim as one of the world's dinosaur capitals. It is built on the Morrison Formation, surrounded by beautiful hills carved out of muted green, red, and gray rocks bursting with dinosaurs. So many dinosaurs have been found here that it's hard to keep track, but the count is probably well over a hundred skeletons by now.

As we drove west from Sheridan, on a surprisingly treacherous road across the rugged Bighorns, I felt I was on the trail of giants. Some of the biggest dinosaurs of all have been found in the Shell area: long-necked sauropods like *Brontosaurus* and *Brachiosaurus*, and the huge carnivores, like *Allosaurus*, that ate them. But I also felt I was walking in the footsteps of another type of giant: the explorers who found the first bones in this area in the late nineteenth century, the railwaymen and laborers who started a dinosaur rush and seized the moment to reinvent themselves as mercenary fossil collectors on the payrolls of gilded institutions like Yale University. They were a ragtag bunch, Wild West ruffians with cowboy hats, mustaches, and unkempt hair, who hacked giant bones out of the ground for months on end, and spent their free time raiding one another's sites, constantly feuding and sabotaging and drinking and shooting. But these unlikely characters revealed a prehistoric world that nobody knew existed.

The first Morrison fossils were surely noticed by the many Native American tribes scattered across the West, but the first recorded bones were collected by a surveying expedition in 1859. In March 1877 the real fun started. A railroad worker named William Reed was returning home from a successful hunt, rifle and pronghorn antelope carcass in tow, when he noticed some huge bones protruding out of a long ridge called Como Bluff, not too far from the railroad tracks in an anonymous expanse of southeastern Wyoming. He didn't know it, but at the same time a college student, Oramel Lucas, was finding similar bones a few hundred miles to the south, in Garden Park, Colorado. That same month, a schoolteacher named Arthur Lakes had just found a cache of fossils near Denver. By the end of that March, the fever of discovery was spreading throughout the American West, to even the most remote villages and railway outposts.

Like any prospecting rush, the dinosaur frenzy attracted a horde of questionable characters to the Wyoming and Colorado backcountry. Many of these men were grizzled opportunists on one mission: to convert dinosaur bones into cash. It didn't take long for them to realize who was paying top dollar: two dapper East Coast academics, Edward Drinker Cope of Philadelphia and Othniel Charles Marsh of Yale University, the same men we briefly met two chapters ago, who studied some of the first Triassic dinosaurs found in western North America. Once chummy, these two scientists had let ego and pride metastasize into a full-on feud, which was so radioactive that they would do anything to one-up each other in an insane battle to see who could name the most new dinosaurs. Cope and Marsh were opportunists, too, and with each letter from a ranch hand or railway porter reporting more new dinosaur bones from the Morrison badlands, they saw the opportunity they had been craving but had been unable to yet fulfill: a chance to beat the other guy once and for all. And they both went for it.

Cope and Marsh treated the West like a battlefield, employing rival teams that often acted more like armies, scooping up fossils wherever they went and sabotaging the other side whenever they could. Loyalties were fluid. Lucas worked for Cope, and Lakes teamed up with Marsh. Reed worked for Marsh, but members

Edward Drinker Cope, the Bone Wars protagonist. *AMNH Library.*

of his team defected to Cope. Pillaging, poaching, and bribing were the rules of the game. The madness continued for over a decade, and when it was over, it was hard to separate the winners from the losers. On the plus side, the so-called Bone Wars led to the discovery of some of the most celebrated dinosaurs, the ones that roll off the tongue of every schoolchild: *Allosaurus*, *Apatosaurus*, *Brontosaurus*, *Ceratosaurus*, *Diplodocus*, *Stegosaurus*, just to name a few. On the other hand, the mentality of constant warfare caused a lot of sloppiness: fossils haphazardly excavated and hastily studied, scraps of bone mistakenly christened as new species, different bits of the skeleton of the same dinosaur regarded as belonging to totally different animals.

A page from Cope's 1874 field notebook, depicting the fossil-rich rocks of New Mexico. *AMNH Library.*

Cope's sketch of a horned dinosaur (ceratopsian), from 1889, an insight into how he envisioned dinosaurs as living animals. (He was much better as a scientist than as an artist.) *AMNH Library.*

Cope's Bone Wars rival, Othniel Charles Marsh (center in the back row), and his team of student volunteers on their 1872 expedition to the American West. *Courtesy of the Peabody Museum of Natural History, Yale University.*

Stegosaurus, one of the most famous dinosaurs discovered in the Morrison Formation during the Bone Wars period. Skeleton on display at the Natural History Museum in London. *PLoS ONE.*

Wars can't last forever, and as the nineteenth century turned over to the twentieth, sanity began to set in. New dinosaurs were still being found throughout the western United States, and most of the country's leading natural history museums and many top universities had crews working somewhere in the Morrison Formation, but the chaos of the dinosaur rush was over. With less turbulence came several major discoveries: A graveyard of over 120 dinosaurs near the Colorado-Utah border, which later became Dinosaur National Monument. A pit with over ten thousand bones, mostly belonging to the superpredator *Allosaurus*, south of Price, Utah, called the Cleveland-Lloyd Dinosaur Quarry. A bone bed in the Oklahoma Panhandle discovered by a road crew and excavated by a team of laborers who lost their jobs during the Great Depression and were put back to work digging up dinosaurs with money from Roosevelt's New Deal. And the site near Shell that Paul Sereno was now working, with the assistance of me and a phalanx of undergraduate laborers paying hefty tuition for the privilege.

Paul has discovered his fair share of dinosaur sites around the world, but the quarry near Shell is not one of them. Instead, it was a local rock collector who reported the first bones in the area. In 1932 she mentioned them to Barnum Brown, a New York paleontologist passing through town. We'll meet Brown again in the next chapter, because much earlier in his career he discovered *Tyrannosaurus rex*. Brown was intrigued by the rock collector's story and followed her to the lonely ranch of an octogenarian named Barker Howe, surrounded by sage-scented hills stalked by mountain lions and abuzz with grazing pronghorns. Brown liked what he saw and stayed the week. What he found was promising enough to convince Sinclair Oil to fund a full-scale expedition in the summer of 1934, to dig up what is now called the Howe Quarry.

It turned out to be one of the most fantastic dinosaur excavations of all time. Once Brown's crew starting digging, they kept finding skeletons everywhere, piled on top of each other and extending in all directions. More than twenty skeletons and four thousand bones in all, covering some 3,000 square feet (280 square meters), approaching the size of a basketball court. There was so much raw fossil material that it took about six months of daily work to excavate; the team broke camp only in mid-November, after enduring two months of heavy snow. The diggers found an entire ecosystem preserved in stone: there were giant long-necked plant-eaters like *Diplodocus* and *Barosaurus*, entangled with sharp-toothed *Allosaurus*es and smaller herbivores that walked on two legs, called *Camptosaurus*. Something horrible had happened here some 155 million years ago. Judging from the contorted angles of their skeletons, the deaths of these animals were neither quick nor painless. Some of the sauropods were found upright, their heavy legs standing tall like columns, stuck in the ancient mud. It seems that these dinosaurs survived a flood, but then were mired in the muck when they tried to run away after the waters receded.

Brown was delighted. He called the site an "an absolute, knockout dinosaur treasure trove!" and gleefully took his haul of dinosaurs back to New York, where they became crown jewels in the collection of the American Museum of Natural History. And then for many decades the Howe Quarry lay dormant, until a fossil collector from Switzerland named Kirby Siber rolled into Wyoming in the late 1980s.

Siber is a commercial paleontologist: he digs up dinosaurs

and sells them. It's a thorny issue for many academic paleontologists like me, who see fossils as irreplaceable natural heritage that should be protected in museums, where they can be studied by researchers and enjoyed by the public, not sold off to the highest bidder. But there is a whole spectrum of commercial paleontologists, ranging from gun-toting criminals who illegally export fossils to diligent, conscientious, well-trained collectors whose knowledge and experience rival that of academics. Siber is in this latter category. In fact, he's the archetype of this kind of collector. He is well respected by researchers and even founded his own dinosaur museum east of Zurich, called the Saurier Museum, which has some of the most remarkable dinosaur exhibits in Europe.

Siber arranged access to the old Howe Quarry but didn't find many dinosaurs. Brown's team had pretty much cleared them all out. So the Swiss collector began prospecting in the surrounding gullies and hills, looking for new sites. It wasn't long before he found a good one, about a thousand feet north of the original quarry. His backhoe first revealed some sauropod bones, and then a string of vertebrae from the backbone of a big, meat-eating theropod. Siber followed the spool-shaped bones, one by one, and before long it was clear that he had something special: the nearly complete skeleton of an *Allosaurus*, the top predator of the Morrison Formation ecosystem. It looked to be the single best fossil of this well-known dinosaur that had ever been found, more than 120 years after Marsh first named it during the heat of the Bone Wars.

Allosaurus was the Butcher of the Jurassic, both figuratively and literally. This fierce predator stalked the Morrison floodplains and riverbanks—think *T. rex* but a little smaller and lighter, about two to two and a half tons in weight and thirty feet (nine meters) long as an adult, and better equipped for running. But it truly earned the title of Butcher, because paleontologists think it used its head like a hatchet to hack its prey to death. Computer models find that the thin teeth of *Allosaurus* couldn't bite very strongly on their own, but the skull could withstand massive amounts of impact force. We also know that *Allosaurus* could open its jaws obscenely wide, so we think a hungry *Allosaurus* would attack with mouth agape and slash down at its prey, slicing through the skin and muscle with its thin but sharp teeth, which were lined up along its jaws like the blades of scissors. Many a *Stegosaurus* and *Brontosaurus* probably breathed its last this way. If for some reason the blood-lusting *Allosaurus* couldn't quite knock off its victim using only its jaws of death, it could always finish the job with a couple of swipes of its clawed, three-fingered arms, which were longer and more versatile than the stubby little forelimbs of *T. rex*.

Finding such a complete and well-preserved *Allosaurus* was one of the highlights of Siber's career, but emotions were about to turn. After the summer's excavation had ended, while Siber was at a fossil show peddling his wares and the *Allosaurus* skeleton remained in the ground, an agent from the US Bureau of Land Management (BLM) happened to be flying over that dusty stretch of northern Wyoming near the Howe Quarry. The agent was checking for signs of fire, part of his job monitoring public lands administered by the US government. But as he glided high above the badlands, he noticed that the dirt roads around the Howe Quarry had been shredded by tire marks. Somebody had been doing some heavy work that summer. That's no problem around the Howe Quarry itself—it's on private land, and Siber had the permission of the landowner. But the BLM agent wasn't quite sure what was private ground and what was public, which could only be worked on by accredited scientists with BLM permission. So he doubled-checked and found that Siber had strayed a few hundred feet into BLM territory. Because Siber had no right to work there, he could no longer excavate the *Allosaurus* skeleton. It was probably an honest mistake, but a costly one.

The BLM now had a problem. A gorgeous dinosaur skeleton was sitting in the ground, and the people who had found it and begun to excavate it couldn't finish the job. So the agency assembled a crack team led by legendary paleontologist Jack Horner's crew at Montana's Museum of the Rockies (Horner is best known for two things: discovering the first dinosaur nesting sites in the 1970s and being the science advisor for the *Jurassic Park* films). Under the eye of television cameras and a throng of newspaper reporters, the academics took out the skeleton and trucked it up to Montana to be carefully conserved in the safety of the laboratory. The dinosaur turned out to be more spectacular than Siber could even have imagined. About 95 percent of the bones were there, an almost unheard of number for a large predatory dinosaur. At about twenty-five feet (eight meters) long, this *Allosaurus* was only about 60 to 70 percent grown. It was still a teenager, but it had already lived a tough life. Its body was covered with all types of maladies: broken, infected, and deformed bones that testify to the rough-and-tumble world of the Late Jurassic, when even the biggest predators didn't have an easy time hunting behemoths like *Diplodocus* and *Brontosaurus*, when the sharpest teeth and claws were no guarantee of surviving a whack from the spiky tail of a *Stegosaurus*.

The *Allosaurus* was nicknamed Big Al, and it became a celebrity dinosaur. It even had its own television special broadcast internationally by the BBC. But once the buzz died down, there remained a huge hole in the ground that was still full of all kinds of fossils that were buried underneath Big Al. Paul Sereno received permission from the BLM to use the site as a field laboratory to teach excavation techniques to his students, and that's why we were taking three big SUVs full of undergraduates there.

During that first season in Wyoming, in the summer of 2005, I spent many days parked out on the high desert, carefully removing globs of popcorn-textured mudstone to help the team uncover the skeleton of a *Camarasaurus*. It may not be one of the brand-name dinosaurs, but *Camarasaurus* is one of the more common species in the Morrison Formation. It is yet another type of sauropod, a close cousin of *Brontosaurus*, *Brachiosaurus*, and *Diplodocus*. *Camarasaurus* had the usual sauropod body: long neck that could reach several stories into the trees, small head with chisel-shaped teeth for stripping leaves, a massive frame that was about fifty feet (fifteen meters) long and weighed around twenty tons. It was probably the type of tasty plant-guzzler that Big Al and the other *Allosaurus*es liked to eat, although its freakish size would have afforded it quite a lot of protection from even the scariest flesh-eaters. Maybe it was a *Camarasaurus* like this one that gave Big Al some of those nasty injuries.

Camarasaurus is one of many enormous sauropods that have been found in the Morrison Formation. It's joined by its famous cousins, the big three of *Brontosaurus*, *Brachiosaurus*, and *Diplodocus*. Then there are the under-the-radar players known only to the cognoscenti (or, perhaps, your average dinosaur-obsessed

The skulls of *Diplodocus* (left) and *Camarasaurus* (right), two sauropods that used their differently shaped skulls and teeth to feed on different types of plants. *Courtesy of Larry Witmer.*

kindergartner): *Apatosaurus*, *Barosaurus*, and further on down the roster, *Galeamopus*, *Kaatedocus*, *Dyslocosaurus*, *Haplocanthosaurus*, and *Suuwassea*. There are various other sauropods that have been named based on scrappy bones, which may belong to even more species. Now, the Morrison Formation covers a wide swath of time, and was deposited across a huge geographic area. Not all of these sauropods lived together. But many of them did—they have been found at the same sites, their skeletons mingled together. The normal situation in the Morrison world was numerous varieties of sauropods cohabiting in the river valleys, their heavy footsteps thundering as they trawled the land in search of the daily hundreds of pounds of leaves and stems that sustained them.

What a weird scene to conjure up! It's akin to imaging five or six different species of elephants crowded onto the African savannahs, all trying to find enough food to survive while lions and hyenas lurk in the background. The Morrison world was no less dangerous. If a sauropod was staggering around with an empty belly, then you could confidently bet an *Allosaurus* was hiding in the brush, ready to pounce at the long-neck's moment of weakness.

In addition to *Allosaurus*, there were many other predators below it on the food chain. There was *Ceratosaurus*, a twenty-foot-long mid-tier hunter with a frightening horn on its snout, a horse-size carnivore named *Marshosaurus* after the Bone Wars pugilist, and a donkey-size primitive cousin of *T. rex* called *Stokesosaurus*. Then you had the slashers: a number of lightly built, fast-running pests like *Coelurus*, *Ornitholestes*, and *Tanycolagreus*, the Morrison version of cheetahs. And all of these meat-gobblers, even *Allosaurus*, probably lived in fear of another monster that reigned near the top of the food chain. It's called *Torvosaurus*, and we don't know much about it, because its fossils are very rare. But the bones we have paint a terrifying picture: a knife-toothed apex predator that was thirty feet (ten meters) long and weighed about two and a half tons or perhaps more, not too far off from the proportions of some of the big tyrannosaurs that would evolve much later.

It's easy to understand why so many predators stalked the Morrison ecosystem: there were a lot of sauropods to eat. It's much more difficult to explain how so many of these giant sauropods lived together. It's an even greater puzzle because there were also plenty of other, smaller plant-eaters that feasted on shrubs closer to the ground: the plate-backed *Stegosaurus* and *Hesperosaurus*, the tanklike ankylosaurs *Mymoorapelta* and *Gargoyleosaurus*, the ornithischian *Camptosaurus*, and a whole zoo of small, fast-running fern-chewers like *Drinker*, *Othnielia*, *Othnielosaurus*, and *Dryosaurus*. The sauropods were also sharing space with all of these herbivores.

So how did the sauropods do it? It turns out that their diversity was their key to success. There were many species of sauropods, yes, but they were all slightly different. Some were absolute colossuses: *Brachiosaurus* was around fifty-five tons, and *Brontosaurus* and *Apatosaurus* tipped the scales in the thirty-to-forty-ton range. But others were smaller: *Diplodocus* and *Barosaurus* were skinny little things, at least as far as sauropods go, weighing a mere ten to fifteen tons. So it goes without saying that some species would need more food than others. These sauropods also had different types of necks: that of *Brachiosaurus* arched proudly into the heavens with the erect profile of a giraffe, perfect for reaching the highest leaves, but *Diplodocus* may not have been able to lift its neck much past its shoulders and may have acted more like a vacuum cleaner sucking up shorter trees and shrubs. Finally, the heads and teeth of these sauropods differed as well. *Brachiosaurus* and *Camarasaurus* had deep, muscle-wrapped skulls and jaws lined with spatula-shaped teeth, so they could eat harder foods like thick stems and waxy leaves. But *Diplodocus* had a long head made up of delicate bones, with a row of tiny pencil-shaped teeth at the front of its snout. It would break its teeth if it tried to eat anything too hard. Instead, it spent its time stripping smaller leaves from the branches, its head rocking back and forth like a rake.

Different species of sauropods were specialized for eating many different kinds of foods—and they had a lot to choose from, as the lush Jurassic forests were cluttered with towering conifers, with thickets of ferns, cycads, and other shrubs down below. The sauropods weren't competing for the same plants, but dividing the resources among themselves. The scientific term for this is niche partitioning—when coexisting species avoid competing with each other by behaving or feeding in slightly different ways. The Morrison world was highly partitioned, which is a sign of just how successful these dinosaurs were. They were carving up almost every square inch of the ecosystem, a dizzying array of species flourishing alongside each other in the hot, humid, waterlogged forests and coastal plains of ancient North America.

But what about Late Jurassic dinosaurs in other parts of the world? The story seems to be the same nearly everywhere we look. We also see a similar cast of diverse sauropods, smaller plant-eating stegosaurs, and small to large carnivores of the *Ceratosaurus* and *Allosaurus* mold in those other places with rich records of Late Jurassic fossils, like China, eastern Africa, and Portugal.

It all comes down to geography. Pangea had started to break up many tens of millions of years before, but it takes a long time for a supercontinent to split. Landmasses can move apart from each other by only a few centimeters each year, about the same pace that our fingernails grow. Thus, there were still big land connections between most parts of the world persisting into the latest Jurassic. Europe and Asia were still globbed together, and they were linked to North America by a series of islands that could easily be traversed by a wayfaring dinosaur. These northern lands—called Laurasia—were beginning to split from southern Pangea, called Gondwana, which was a stuck-together mess of Australia, Antarctica, Africa, South America, India, and Madagascar. Laurasia and Gondwana were intermittently connected by land bridges when sea level was low, and even during times of higher water, other islands provided a convenient migratory route between north and south.

The Late Jurassic, then, was a time of global uniformity. The same suite of dinosaurs ruled every corner of the globe. Majestic sauropods divided food among them, reaching a peak of diversity unmatched by any other large plant-eaters in Earth history. Smaller plant-chewers prospered in their shadows, and a motley crew of meat-eaters took advantage of all of that herbivore flesh. Some, like *Allosaurus* and *Torvosaurus*, were the first truly giant theropods. Others, like *Ornitholestes*, were the founding members of that dynasty that would eventually produce *Velociraptor* and birds. The planet was sweltering and the dinosaurs were able to move around wherever they wanted. This was the real Jurassic Park.

145 **MILLION YEARS** ago, the Jurassic Period transitioned into the final stage of dinosaur evolution, the Cretaceous Period. Sometimes the switch between geological periods happens with a flourish, as when the megavolcanoes closed out the Triassic. Other times, it's barely noticeable, and more a matter of scientific bookkeeping, a way for geologists to break up long stretches of time without any major changes or catastrophes. The changeover between the Jurassic and Cretaceous is that type of boundary. There was no calamity like an asteroid impact or a big eruption that ended the Jurassic, no sudden die-off of plants and animals, no brave new world as the Cretaceous dawned. Rather, the clock just ticked over, and the diverse Jurassic ecosystems of giant sauropods, plate-backed dinosaurs, and small to monstrous meat-eaters continued into the Cretaceous.

That's not to say, however, that *nothing* changed, for plenty was happening to the Earth around the Jurassic-Cretaceous boundary—no apocalyptic disasters but slower changes to the continents, oceans, and climate occurring over some 25 million years. The hothouse world of the Late Jurassic was interrupted by a cold snap, followed by a turn to more arid conditions, before things swung back to normal in the Early Cretaceous. Sea levels started to fall during the latest Jurassic and stayed low across the boundary, until the waters started to rise again some 10 million years into the Cretaceous. With low sea levels came a lot more exposed land, which allowed dinosaurs and other animals to move around even more easily than during the Late Jurassic. Pangea continued to rupture, the fragments of the supercontinent moving farther and farther apart from one another as time marched on. Gondwana, that huge expanse of southern lands, finally began to split, the cracks starting to define the shapes of today's southern hemisphere continents. First the conjoined mass of Africa and South America detached from the section of Gondwana containing Antarctica and Australia, and then this latter chunk also began to fracture. Volcanoes welled up through the fissures, and although none were on the scale of the monster eruptions at the end of the Permian or Triassic, they would have brought with them the same nasty stew of environment-poisoning lava and gases.

None of these changes were particularly deadly on their own, but together they were an insidious cocktail of dangers. The long-term shifts in temperature and sea level were probably unrecognizable to dinosaurs, the sort of thing that neither they nor any of us, had we been around, would ever have noticed in one lifetime. Plus, in the dinosaur-eat-dinosaur world of the Late Jurassic and Early Cretaceous, the *Brontosaurus*es and *Allosaurus*es had more important things to stress over than little changes in the tide line or slightly cooler winters. Given enough time, however, these changes built up and became silent killers.

By about 125 million years ago, some 20 million years after the Jurassic ended, a new Cretaceous world had emerged, ruled by a very different suite of dinosaurs. The most obvious change had to do with the most prominent dinosaurs—the gargantuan sauropods. Once so diverse in the Late Jurassic Morrison ecosystems, the long-necks suffered a crash in the Early Cretaceous. Almost all of the familiar species like *Brontosaurus*, *Diplodocus*, and *Brachiosaurus* went extinct, while a new subgroup called the titanosaurs began to proliferate, eventually evolving into supergiants like the middle Cretaceous *Argentinosaurus*, which at more than a hundred feet (thirty meters) long and fifty tons in mass was the largest animal known to have ever lived on land. But despite the outlandish sizes of the new Cretaceous species, never again would sauropods be as dominant as they were in the Late Jurassic; never again would they boast such a variety of necks and skulls and teeth that allowed them to exploit so many ecological niches.

As sauropods suffered, smaller plant-eating ornithischians blossomed, becoming ubiquitous midsize herbivores in ecosystems around the world. Most famous of these is surely *Iguanodon*, one of the very first fossils to be called a dinosaur, after it was discovered in the 1820s in England. *Iguanodon* was about thirty feet (ten meters) long and weighed a few tons. It had a spike on its thumb for defense and a beak at the front of its mouth for snipping plants, and it could switch between walking on all fours and sprinting on its hind legs. Its line would eventually go on to produce the hadrosaurs, or duck-billed dinosaurs, the group of amazingly successful herbivores that thrived at the very end of the Cretaceous, alongside their nemesis, *T. rex*. That was still many tens of millions of years in the future, but those seeds were planted in the Early Cretaceous.

While the iguanodons were stepping in for the smaller sauropods, there were also changes afoot among ground-feeding herbivores. The plate-backed stegosaurs went into long-term decline, gradually wasting away until the last surviving species succumbed to extinction sometime in the Early Cretaceous, snuffing out this iconic group once and for all. Replacing them were the ankylosaurs, freakish creatures whose skeletons were covered in armor plates, like a reptilian Panzer. They had originated back in the Jurassic and remained marginal understudies in most ecosystems, but they exploded in diversity as stegosaurs regressed. Ankylosaurs were some of the slowest, stupidest dinosaurs of all, but they made a living happily chomping ferns and other low-lying vegetation, their body armor making them impervious to attack. Not even the sharpest-toothed predator could get in a good chomp when it had to bite through several inches of solid bone.

Then there were the meat-eaters. With so much going on with their herbivore prey, it's no surprise that theropods experienced their own drama as the Jurassic turned into the Cretaceous. A much greater diversity of small carnivores appeared, and some of them started to experiment with weird diets, trading in meat for nuts, seeds, bugs, and shellfish. One group, the scythe-clawed therizinosaurs, even went full vegetarian. On the other end of the size spectrum, a weird clan of large theropods called spinosaurids evolved sails on their backs and long snouts full of cone-shaped teeth, and moved into the water, where they started behaving like crocodiles and eating fish.

However, as is usually the case when it comes to theropods, the most gripping story line concerns the apex predators. Like their smaller brethren, the top-of-the-food-chain supercarnivores also experienced massive upheaval across the Jurassic-Cretaceous boundary. These species are some of my favorites, because the very first dinosaurs that I studied—as an undergraduate working with Paul Sereno, during those same summers we dug up Late Jurassic sauropods in Wyoming—were giant theropods from the Early Cretaceous of Africa.

WHEN I WAS a teenager, I watched movies and listened to music and went to baseball games—the normal stuff—but my hero wasn't some athlete or actor. He was a paleontologist. Paul Sereno, the National Geographic Explorer in Residence, dinosaur hunter extraordinaire, leader of expeditions all over the world, and one of *People* magazine's 50 Most Beautiful People, in the issue with Tom Cruise on the cover. I was a dinosaur-obsessed high schooler, and I followed Sereno's work like a rock star's groupie. He taught at the University of Chicago, not too far from where I lived, and he grew up in Naperville, Illinois, a suburb where some of my cousins were from. He was a local kid who did good, who became a celebrity scientist and adventurer, and I wanted to be like him.

I met my hero when I was fifteen years old, when he was lecturing at a local museum. I'm sure Paul was used to meeting fanboys, but I upped the weird factor when I shoved a manila envelope in his face, so full of photocopied magazine pages that it couldn't be sealed shut. You see, I was also a budding journalist at the time, or at least I thought I was, and I was churning out articles for amateur paleontology mags and websites at a pace that bordered on the creepy. Many of them were about Paul and his discoveries, and I wanted him to see the things I had written about him. My voice cracked as I handed him the envelope. It was awkward. But Paul was very nice to me that afternoon, and after a long chat, he told me to keep in touch. I met him a few more times over the next couple of years. We exchanged a lot of e-mails, and when I decided to put journalism aside and dive into paleontology as a career, there was only one college that I wanted to attend: the University of Chicago, so I could learn under Paul.

Chicago accepted my application, and I enrolled in the autumn of 2002. During freshman week, I met with Paul and begged him to let me work in his basement fossil lab, where his newest treasures from Africa and China were being revealed, entirely new dinosaurs coming into focus as sand grains were scrubbed away from the bones. I would do anything—even wash the floors or clean the shelves. Thankfully, Paul channeled my enthusiasm elsewhere. He began by teaching me how to conserve and catalog fossils, and then one day he had a surprise. "How would you like to describe a new species of dinosaur?" he asked while leading me to a row of cabinets.

Spread in front of me, in drawer after drawer, were fossils of Early to middle Cretaceous dinosaurs that Paul and his team had recently brought back from the Sahara Desert. About a decade earlier, after concluding his wildly successful expeditions to Argentina that netted the primitive dinosaurs *Herrerasaurus* and *Eoraptor*, Paul shifted his attention to northern Africa. At the time, little was known about African dinosaurs. A few excursions led by Europeans during the colonial period had found some intriguing fossils in places like Tanzania, Egypt, and Niger, but once the colonizers left, so too did most interest in collecting dinosaurs. Not only that, but some of the most important African collections—made by the German aristocrat Ernst Stromer von Reichenbach, from the Early to mid Cretaceous rocks of Egypt—weren't around anymore. They had the great misfortune of being kept in a museum just a few blocks from Nazi headquarters in Munich and were destroyed by an Allied bombing raid in 1944.

When Paul turned his focus to Africa, all he had to go by were some photographs, a few published reports, and a smattering of bones in those European museums that weren't blitzed during the war. That didn't stop him, though. He mounted a reconnaissance trip to Niger, in the heart of the Sahara, in 1990. His team found so many fossils that they returned again in 1993, 1997, and several times after that. These were arduous trips—proper Indiana Jones–style expeditions, often lasting for several months and afflicted by the occasional bandit attack or civil war. As something of a break, they took a year off and visited Morocco in 1995. There, too, they uncovered a bounty of bones, including the gorgeously preserved skull of a giant flesh-eater called *Carcharodontosaurus*, a dinosaur that Stromer had originally named based on a partial skull and skeleton from Egypt that were among those fossils incinerated in the Munich museum. All told, Paul's African expeditions collected some one hundred tons of dinosaur bones, many of which still sit in a warehouse in Chicago, waiting to be studied.

Those dinosaurs not in the warehouse are inventoried at Paul's lab, and these were the bones laid out before me. Some belonged to a weird sauropod called *Nigersaurus*, a plant-inhaling machine with hundreds of teeth packed into the front edge of its jaws. There were several elongate vertebrae of the fish-eating spinosaurid *Suchomimus*—the bones that supported the tall sail that extended along its back. Nearby was the gnarly-textured skull of a carnivore called *Rugops*, which probably scavenged carcasses as much as it hunted.

And these fossils weren't only dinosaurs. There was the man-size cranium of the forty-foot-long crocodilian *Sarcosuchus*—appropriately nicknamed SuperCroc by the media-savvy Sereno—and the wing bones of a large pterosaur, and even some turtles and fish. All of these fossils came from rocks that formed over some ten to fifteen million years of the Early to middle Cretaceous, in river deltas and along the shores of warm tropical seas fringed by mangrove forests, back when the Sahara was a steamy swampy jungle instead of a desert.

As my eyes darted among the fossils, the cast of characters expanding as each drawer opened, Paul stopped and picked up a bone. It was part of the face of a huge-meat eating dinosaur that looked to be almost as big as *T. rex*. There were other things in the same drawer: a piece of a lower jaw, some teeth, and a fused mass of bones from the back of the skull, which would have surrounded the brain and ears. Paul recounted how he'd discovered the specimens a few years back in a desolate part of Niger called Iguidi, just west of a desert oasis, in red sandstones left by a river between 100 and 95 million years ago. He could tell that they were similar to the *Carcharodontosaurus* bones that he'd collected in Morocco, but the match wasn't perfect. He wanted me to figure out the discrepancy.

I was nineteen years old, and this was my first taste of the detective work that goes into identifying dinosaurs. It was intoxicating. I spent the remainder of the summer scrutinizing the bones, measuring and photographing them, comparing them with other dinosaurs. I concluded that the bones from Niger were indeed very similar to the Moroccan skull of the species *Carcharodontosaurus saharicus* but found that there were also so many differences that the two could not have belonged to the same species. Paul agreed, and we wrote up a scientific paper describing the Niger fossils as a new dinosaur, a close but distinct relative of the Moroccan species. We called it *Carcharodontosaurus iguidensis*. It was the alpha predator of those humid seaside ecosystems of mid-Cretaceous Africa, a forty-foot-long, three-ton beast that lorded over all of the other dinosaurs that Paul had been trawling out of the Sahara.

There was a whole group of dinosaurs like *Carcharodontosaurus* that lived throughout the world during the Early to middle Cretaceous. They are named, perhaps unoriginally, the carcharodontosaurs. Among the family album are three species—*Giganotosaurus*, *Mapusaurus*, and the hauntingly named *Tyrannotitan*—all from South America, which during the Early to middle Cretaceous was still connected to Africa. Other siblings lived farther afield: *Acrocanthosaurus* in North America, *Shaochilong* and *Kelmayisaurus* in Asia, and *Concavenator* in Europe. And there's also another one from the Sahara, called *Eocarcharia*, which Paul and I described based on some skull bones he found on another expedition to Niger. It was about 10 million years older than *Carcharodontosaurus*, and only about half the size. It was about as brutish as a dinosaur could get, with a gnarled knob of bone and skin above each eye that gave it an evil scowl and may have even been used to head-butt prey into submission.

These carcharodontosaurs fascinated me. They were basically doing what tyrannosaurs would do many tens of millions of years later: supersizing their bodies, developing an arsenal of predatory weapons, and terrorizing every living thing from their undisputed perch at the top of the food pyramid. Where did they come from? How did they spread around the world and become so dominant? And then what happened to them?

There was only one way to answer these questions. I needed to build a family tree. Genealogy is a key to understanding history, which is why so many people, me included, are obsessed with our own family trees. Knowing the connections among kin helps to untangle how our families have changed over the centuries: when and where our ancestors lived, when a migration or an unexpected death occurred, how the family merged with others through marriages. The same with dinosaurs. If we can read their family tree—or their phylogeny, as paleontologists technically call it—we can use it to illuminate their evolution. But how do you make a family tree for dinosaurs? *Carcharodontosaurus* doesn't have a birth certificate, and the ancestor of *Giganotosaurus* wasn't granted a visa when it left Africa for South America. But there is another type of clue coded in the fossils themselves.

Evolution causes change over time, particularly in the appearances of organisms. When two species diverge from each other, usually only minor differences separate them, and you may have a hard time telling them apart at a glance, but as time ticks on and the two lineages go their separate ways, they become more and more different from each other. It's the same reason that I look a lot like my father but barely resemble my third cousins. The other thing evolution occasionally does is produce new things—an extra tooth, or a horn sticking out of the forehead, or a mutation that causes a finger to be lost. These novelties will be inherited by the descendants of the first critter to develop them, but they won't be seen in cousins that had already split off and started evolving down their own path. I've inherited all kinds of things from my parents, and my children will then inherit those things from me. But if my cousin suddenly goes weird and grows a set of wings, they can't be passed on to me, because there is no direct line of descent between us. That means, thankfully in this case, that none of my children get those wings either.

Genealogy, therefore, is written into the way we look. On the whole, dinosaurs with similar skeletons are probably more closely related to each other than to other species that look drastically different. But if you want to know if two dinosaurs truly are close brethren, you need to look out for those evolutionary novelties, because animals that possess a newly evolved feature like an extra finger must be more closely related to each other than to ones that don't have it. That's because they must have inherited that novelty from a common ancestor, which developed the feature and started an evolutionary domino effect of passing it down the bloodline, generation by generation. Any species with that extra finger is part of the bloodline; anything without it is likely on another side branch of the family tree. So to build a genealogy of dinosaurs we need to pore over their bones, find a way to assess how similar and different they are, and identify evolutionary novelties and which subsets of the dinosaurs in question share them.

When I became intrigued by carcharodontosaurs, I began to track down as much information on each species as I could. I visited museums to study skeletons firsthand, and I gathered photographs, drawings, published literature, and notes for some of the more exotic fossils in faraway places that were inaccessible to an unfunded undergraduate. The more I looked, I recognized features of the bones that varied among species. Some carcharodontosaurs had deep sinuses surrounding their brain, others did not. The giant ones like *Carcharodontosaurus* had massive, bladelike teeth that kind of resembled those of sharks (hence its name, which means "shark-toothed lizard"), but the smaller species had much daintier chompers. The list went on and on, until I had come up with ninety-nine different ways that some of these predators differed from others.

Now it was time to make some sense out of this information. I turned my list into a spreadsheet: each row a species, each column one of the features of the anatomy, each data cell filled with a 0, 1, or 2 denoting the different versions of each feature seen in that species. Dainty teeth in *Eocarcharia*, 0; sharky teeth in *Carcharodontosaurus*, 1. Then I opened the spreadsheet in a computer program that uses algorithms to search through the maze of data and generate a family tree. It pinpoints which anatomical features are novelties and then identifies which species share them. This may sound trivial, but the computer is necessary because the distribution of novelties can be complicated. Some are seen in many species—those big sinuses around the brain are present in most carcharodontosaurs. Others are much rarer, like the shark-mimic teeth, which are seen only in *Carcharodontosaurus*, *Giganotosaurus*, and their closest kin. The computer is able to take all of this complexity and recognize a Russian doll pattern. If two species share many novelties between only themselves, they must be each other's closest relatives. If those two species share other novelties with a third animal, those three must be more closely related to one another than to the remainder of the dinosaurs. And so on, until a complete family tree has been drawn. This whole process is what we in the business call a cladistic analysis.

My family tree of carcharodontosaurs helped me unravel their evolution. First, it clarified where these colossal carnivores came from and how they rose to glory. They got their start in the Late Jurassic and are very close relatives of that most terrifying predator of the Jurassic, the Butcher itself, *Allosaurus*. In effect, they evolved from a legion of hypercarnivores that was already incumbent in the apex predator niche, and then they escalated things further by becoming larger, stronger, and fiercer when their ancestors went extinct at the end of the Jurassic, 145 million years ago, during that long night of environmental and climate change. Did they drive these other allosaurs to extinction or take advantage when they succumbed for other reasons? We don't yet know the answer. In either event, the carcharodontosaurs found a way to usurp the place of their forebears and, as the Cretaceous dawned, the kingdom was now theirs. For the next 50 million years or so, deep into the middle Cretaceous, the carcharodontosaurs ruled the world.

The genealogy also gives insight into something else: why these flesh-gouging monsters lived where they did. Because they originated when most of the continents were still connected during the Late Jurassic, the first carcharodontosaurs easily spread around the world. As time went on and the continents fragmented further, different species became isolated in different areas. The structure of their family tree shows this—it reflects the motion of the continents. Some of the last carcharodontosaurs to evolve were a clan of South American and African species. (South America and Africa remained connected to each other long after links with North America, Asia, and Europe were severed.) Isolated south of the equator, the members of this clan—*Giganotosaurus*, *Mapusaurus*, and the *Carcharodontosaurus* from Niger that I studied with Sereno—grew to sizes previously unheard of for meat-eating dinosaurs.

Nevertheless, as ferocious as these carcharodontosaurs were, they wouldn't stay on top forever. Living alongside them, in their shadows, was another breed of carnivore. Smaller, faster, brainier. Their name, the tyrannosaurs. They would soon make their move and begin a new dinosaur empire.

5

The TYRANT DINOSAURS

Qianzhousaurus

ONE SWELTERING SUMMER DAY IN 2010, a backhoe operator in the southeastern Chinese city of Ganzhou heard a loud crunch. He expected the worst. His crew was racing to finish an industrial park—a sprawling monotony of offices and warehouses of the sort that I've seen crop up all over China during the past decade. Any delay could be pricey. Maybe he had hit impenetrable bedrock, an old water main, or another nuisance that would stall the project.

When the dirt and smoke cleared, however, he didn't see any mangled pipes or wires. There was no bedrock in sight. Instead, something very different came into focus: fossilized bones, lots of them, some of them enormous.

Construction halted. The workman didn't have any advanced degrees or training in paleontology, but he realized his discovery was important. He knew it must be a dinosaur. His homeland had become the epicenter of new dinosaur discoveries, the place where about half of all new species are being found these days. So he called over his foreman, and that is when the madness began.

This dinosaur had been buried for more than 66 million years, but now its fate was down to the kind of quick decisions that unfold during a crisis. Word started to leak out. In a panic, the foreman called a friend from town, a fossil collector and dinosaur enthusiast known to posterity only as Mr. Xie. Grasping the gravity of the discovery, Mr. Xie—his honorific and hazy surname recalling one of those shadowy characters in a Bond film—raced to the worksite and rang up some mates at the town's mineral resources department, a branch of the local government. The game of telephone continued and the agency was able to round up a small team to gather the bones. It took them six hours, but they collected every scrap they could find. They filled twenty-five bags with dinosaur bits and took them to the town's museum for safekeeping.

Their timing was perfect—ominously so. Just as the team was finishing, three or four fossil traffickers appeared on the scene. Like bloodhounds, these black-market hucksters caught the scent of a new dinosaur and wanted to buy it for themselves. A little bit of bribe money would turn into a major payday if they sold the new dinosaur to some wealthy foreign businessman with a taste for exotic fossils. This kind of thing is all too common in China and in many other parts of the world (although it is often against the law). It is heartbreaking to think of the fossils that have been lost to the dark underworld of illegal dealing and organized crime. But this time, the good guys won.

When scientists examined the fossils in the safety of the local museum and began to piece together the bones, they quickly recognized how incredible this new discovery was. It wasn't just a jumble of random bones but a nearly complete skeleton of a predatory dinosaur, one of the massive, sharp-toothed behemoths that always seem to play the villain in films and television documentaries. And the skeleton looked similar to a famous dinosaur from halfway around the world: the great *Tyrannosaurus rex*, which stalked the forests of North America at about the same time that these red rocks from Ganzhou, which the backhoe operator was plowing through to lay his foundation, were formed.

Then it clicked: they were looking at an Asian tyrannosaur. The ferocious ruler of a 66-million-year-old world of dense jungles, sticky with humidity all year round, with swamps and the occasional quicksand pit peppered in between the ferns, pines, and conifers. It was an ecosystem teeming with lizards, feathered omnivorous dinosaurs, sauropods, and swarms of duck-billed dinosaurs, some of which got caught in the slushy death pits and were preserved as fossils. The ones lucky enough to survive were tasty prey for the creature the workman stumbled upon by pure chance: one of the closest relatives of *T. rex*.

THAT BLESSED WORKMAN. He had made a discovery of the sort most paleontologists dream of. Lucky for me, this was a finding that I got to be part of without having to do the hard work of hunting it myself.

A few years after the craziness of that late summer day in Ganzhou, I was at a conference at the Burpee Museum of Natural History, in the frozen winter wasteland of northern Illinois, just up the road from where I grew up. Scientists from around the world had gathered to discuss the extinction of the dinosaurs. Earlier in the day, I was mesmerized by a presentation from Junchang Lü, my eyes opening wider with each slide, as photo after photo of beautiful new fossils from China flashed across the screen. I knew Professor Lü by reputation. He was widely regarded as one of China's top dinosaur hunters, a man whose discoveries helped establish his country as the world's most exciting place for dinosaur research.

Professor Lü was a star. I was a young researcher, but to my great surprise Professor Lü approached me. I shook his hand and congratulated him on his talk, and we exchanged a few other pleasantries. But there was urgency in his voice, and I noticed he was clutching a folder thick with photos. Something was going on.

Professor Lü told me he had been tasked with studying a

spectacular new dinosaur found by a construction worker in southern China a few years before. He knew it was a tyrannosaur, but it seemed peculiar. It was different enough from *T. rex* that it must be a new species. And it looked kind of similar to a weird tyrannosaur that I had described a few years earlier as a graduate student—a slender, long-snouted predator from Mongolia called *Alioramus*. But Professor Lü wasn't sure. He wanted a second opinion. Of course I offered to help in any way I could.

Professor Lü, or Junchang, as I soon knew him, told me all about his past—how he grew up poor in Shandong Province, on China's eastern coast, a child of the Cultural Revolution who staved off hunger by picking wild vegetables. Then, once the winds of politics changed, he studied geology in college, went to Texas to do his PhD, and came back to Beijing to take up one of the most vaunted jobs in Chinese paleontology, a professorship at the Chinese Academy of Geological Sciences.

Junchang—the peasant turned professor—became my friend. Not too long after we met at the conference, he invited me to China to help him study the new tyrannosaur and write up a scientific paper describing the skeleton. We scrutinized each part of the skeleton, comparing it to all other tyrannosaurs. We confirmed that it was a close cousin of *T. rex*. A little over a year later, in 2014, we unveiled the workman's chance discovery as the newest member of the tyrannosaur family tree, a new species that we called *Qianzhousaurus sinensis*. The formal name is a something of a tongue twister, so we nicknamed it Pinocchio rex, in reference to its funny long snout. The press got wind of the discovery—journalists seemed to love the silly nickname—and Junchang and I were amused to see our faces splashed across the British tabloids the morning after our announcement.

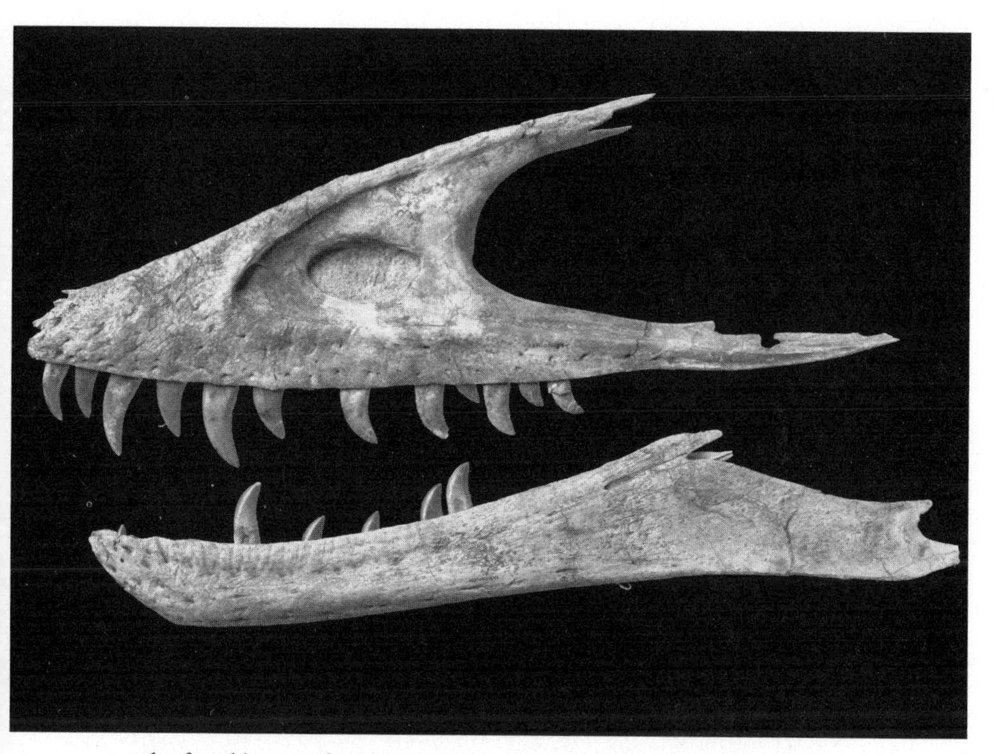

The facial bones of *Alioramus altai*, a new species of long-snouted tyrannosaur from Mongolia that I described as a PhD student. *Photograph by Mick Ellison.*

Qianzhousaurus is part of a surge of new tyrannosaur discoveries over the past decade that is transforming our understanding of this most iconic group of meat-eating dinosaurs. *T. rex* itself has been in the limelight for over a century, since it was first discovered in the early 1900s. It's the king of dinosaurs, a forty-foot-long, seven-ton behemoth on a first-name basis with almost everyone on the planet. Later during the twentieth century, scientists discovered a few close relatives of *T. rex* that were also impressively large and realized that these big predators formed their own branch of dinosaur genealogy, a group that we called the tyrannosaurs (or Tyrannosauroidea in formal scientific parlance). However, paleontologists struggled to understand when these fantastic dinosaurs originated, what they evolved from, and how they were able to grow so large and reach the top of the food chain. These questions have remained unanswered until now.

Over the last fifteen years, researchers have recovered nearly twenty new tyrannosaur species at locations the world over. The dusty southern Chinese construction site that yielded *Qianzhousaurus* is one of the least unusual places where a new tyrannosaur has been found. Other new species have been pried from the sea-battered cliffs of southern England, the frigid snowfields of the Arctic Circle, and the sandy expanses of the Gobi Desert. These finds have allowed my colleagues and me to build a family tree of tyrannosaurs in order to study their evolution.

The results are surprising.

It turns out that tyrannosaurs were an ancient group that originated more than 100 million years before *T. rex*, during those golden days of the Middle Jurassic when dinosaurs were beginning to prosper and long-necked sauropods, like the creatures whose footprints we found in that ancient Scottish lagoon, were rumbling across the land. These first tyrannosaurs weren't very impressive. They were marginal, human-size carnivores. They continued this way for another 80 million years or so, living in the shadows of larger predators, first *Allosaurus* and its kin in the Jurassic, and then the fierce carcharodontosaurs in the Early to middle part of the Cretaceous. Only then, after that interminable period of evolution in anonymity, did tyrannosaurs start growing bigger, stronger, and meaner. They reached the top of the food chain and ruled the world during the final 20 million years of the Age of Dinosaurs.

THE STORY OF tyrannosaurs begins with the discovery of *T. rex*, the namesake of the group, in the early days of the twentieth century. The scientist who studied *T. rex* was a good friend of President Theodore Roosevelt's, a boyhood chum who shared Teddy's love for nature and exploration. His name was Henry Fairfield Osborn, and during the early 1900s, he was one of the most visible scientists in the United States.

Osborn was president of New York's American Museum of Natural History and of the American Academy of Arts and Sciences, and in 1928 he even graced the cover of *Time* magazine. But Osborn was no normal man of science. His blood ran blue: his father was a railroad tycoon, his uncle the corporate raider J. P. Morgan. He seemed to be a member of every wood-paneled, smoke-filled, good-old-boy backroom club there was. When he wasn't measuring fossil bones, he was rubbing shoulders with New York's social elite in the penthouses of the Upper East Side.

Osborn is not remembered very fondly today. He wasn't a very nice man. He used his wealth and political connections to push pet ideas on eugenics and racial superiority. Immigrants, minorities, and the poor were seen as enemies. Once Osborn even organized a scientific expedition to Asia with the hope of finding the very oldest human fossils, to prove that his species couldn't possibly have originated in Africa. He couldn't fathom being the evolutionary descendent of an "inferior" race. No wonder he is often dismissed today as just another bygone bigot.

Osborn is probably not the type of guy I would want to have a beer—or more likely, a really fancy cocktail—with if I found myself in Gilded Age New York. (I speculate, but he might not have sat down with me anyway, leery of my very ethnic-sounding Italian name.) Nevertheless, there's no denying that Osborn was a clever paleontologist and an even better scientific administrator. It was in his capacity as president of the American Museum of Natural History—the august institution that rises like a cathedral on the west side of Central Park, where I would later work on my PhD—that Osborn made one of the best calls in his career. He decided to send a sharp-eyed fossil collector named Barnum Brown out to the American West in search of dinosaurs.

We briefly met Brown in the last chapter, when a much older version of him was excavating Jurassic dinosaurs in the Howe Quarry in Wyoming. He was an unlikely hero. He grew up in a speck of a village on the Kansas prairie, a coal-company town where only a few hundred people lived. Maybe his parents gave him a flamboyant name inspired by the circus showman P. T. Barnum in some attempt to escape the drudgery of their rural life. The young Barnum didn't have many people around to talk to, but he was surrounded by nature, and he became infatuated with rocks and shells. He even started a little museum at his house, something my dinosaur-obsessed younger brother, also growing up in a placid Midwestern town, would later do after seeing *Jurassic Park* in the cinema. Brown went on to study geology in college and then made his way from the small time to New York City in his twenties. It was there he met Osborn and was hired as a field assistant, tasked with bringing huge dinosaurs from the unexamined expanses of Montana and the Dakotas to the bright lights of Manhattan, where socialites who had never slept a night outdoors could gawk at the stupendousness of it all.

Barnum Brown (left) and Henry Fairfield Osborn digging up dinosaur bones in Wyoming, 1897. *AMNH Library.*

This is how Brown found himself, in 1902, in the desolate badlands of eastern Montana. While out prospecting the hills, Brown came across a jumble of bones. Part of a skull and jaw, some vertebrae and ribs, bits of the shoulder and arm, and most of the pelvis. The bones were enormous. The size of the pelvis indicated an animal that stood several meters tall, certainly much larger than a human. And they were clearly the remains of a muscular creature that could run relatively fast on two legs—

the characteristic body type of a meat-eating dinosaur. Other predatory dinosaurs had been found before—like *Allosaurus*, the Butcher of the Late Jurassic—but none of these were anywhere near the colossal size of Brown's new beast. He was on the cusp of turning thirty years old, and he had made a discovery that would define him for the rest of his life.

Brown sent his discovery back to New York, where Osborn was anxiously awaiting the shipment. The bones were so big, they took years to clean up and assemble into a partial skeleton that could be exhibited to the public. This work was mostly done by the end of 1905, when Osborn announced the new dinosaur to the world. He published a formal scientific paper designating the new dinosaur as *Tyrannosaurus rex*—a beautiful combination of Greek and Latin that means "tyrant lizard king"—and put the bones on display at the American Museum, as the institution is known among scientists. The new dinosaur became a sensation, making headlines throughout the country. The *New York Times* celebrated it as "the most formidable fighting animal" that had ever existed. Crowds flocked to the museum, and when they came face-to-face with the tyrant king, they were aghast at its monstrous size and dumbfounded by its ancient age, then estimated at some 8 million years old (we now know that it is much older, about 66 million years old). *T. rex* had become a celebrity, and so had Barnum Brown.

Brown will always be remembered as the man who discovered *Tyrannosaurus rex*, but this was just the start of his career. He developed such an eye for fossils that he steadily progressed from a fossil-collecting grunt worker to the curator of vertebrate paleontology at the American Museum, the scientist in charge of the world's finest dinosaur collection. Today, if you visit its spectacular dinosaur halls, many of the fossils you'll see were collected by Brown and his teams. No wonder that Lowell Dingus, one of my former colleagues in New York who wrote a biography of Brown, refers to him as "the best dinosaur collector who ever lived." This sentiment is shared by many of my fellow paleontologists.

Brown was the first celebrity paleontologist, acclaimed for his lively lectures and a weekly CBS radio show. People would flock to see him as he passed through the American West on trains, and later in his life he helped Walt Disney design the dinosaurs in *Fantasia*. Like any good celebrity, Brown was an eccentric. He hunted fossils in the dead of summer in a full-length fur coat, made extra cash spying for governments and oil companies, and had such a fondness for the ladies that rumors of his tangled web of offspring are still whispered throughout the western American plains. You can't help but think that if Brown were alive today, he would be the star of some outrageous reality show. And probably a politician.

A few years after *T. rex* stormed New York, Brown was back at it, in his fur coat, scrambling over the badlands of Montana, looking for more fossils. As usual, he found them. This time it was a much better *Tyrannosaurus*: a more complete skeleton with a gorgeous skull, nearly as long as a man and with over fifty sharp teeth the size of railroad spikes. While Brown's first *T. rex* was too scrappy to make a good estimate for the total size of the animal, the second fossil showed that rex was a king indeed: a dinosaur well over thirty-five feet long that must have weighed several tons. There was no doubt: *T. rex* was the largest and most fearsome land-living predator that had ever been discovered.

FOR THE NEXT few decades, *T. rex* enjoyed life at the top, the star of movies and museum exhibits around the globe. It battled the giant gorilla in *King Kong* and terrified audiences in the screen adaptation of Arthur Conan Doyle's *The Lost World*. But this fame masked a puzzle: for nearly the entire twentieth century, scientists had little idea of how *T. rex* fit into the broader picture of dinosaur evolution. It was an oddball, a creature so much larger and so dramatically different from other known predatory dinosaurs that it was difficult to place in the dinosaur family album.

During the first few decades after Brown's discovery, paleontologists unearthed a handful of close *T. rex* relatives in North America and Asia. To nobody's surprise, Brown himself made some of the most important of these discoveries, most notably a mass graveyard of big tyrannosaurs in Alberta in 1910. These *T. rex* cousins—*Albertosaurus*, *Gorgosaurus*, *Tarbosaurus*—are quite similar to *T. rex* in size and have nearly identical skeletons. As the science of dating rocks advanced during the later twentieth century, it was also determined these other tyrannosaurs lived at about the same time as *T. rex*: the very latest Cretaceous, between 84 and 66 million years ago. So scientists were in a quandary. There were a bunch of huge tyrannosaurs at the top of the food chain thriving at the peak of dinosaur history. Where did they come from?

That mystery has been answered only very recently, and as with so much of what we've learned about dinosaurs over the last few decades, our new understanding of tyrannosaur evolution stems from a wealth of new fossils. Many of these have come from unexpected locales, perhaps none more so than what is currently recognized as the very oldest tyrannosaur, a modest little critter called *Kileskus* that was discovered in 2010 in Siberia. When you think of dinosaurs, Siberia is probably not a place that comes to mind, but their fossils are now being found throughout the world, even the far northern reaches of Russia, where paleontologists need to cope with harsh winters and humid, mosquito-infested summers.

Alexander Averianov, my friend from the Zoological Institute of the Russian Academy of Sciences in Saint Petersburg, is one of those paleontologists. Sasha, as we all call him, is among the world's experts on those puny mammals that lived alongside (or more correctly, underneath) the dinosaurs. He also studies the dinosaurs that were keeping his beloved mammals down. Sasha began his career as the Soviet Union was disintegrating, and through his numerous discoveries and meticulous descriptions of fossil anatomy, he has now become one of the leading paleontologists in the new Russia.

A few years ago, Sasha showed me a new dinosaur fossil from Uzbekistan at a conference. He whisked me up to his room, ceremoniously opened an ornately colored orange-and-green cardboard box, and pulled out part of the skull of a meat-eater. He put the fossil back in the box and handed it to me so I could take it back to Edinburgh to CAT-scan it. But before he let go, he looked me in the eye and, in the Russian-accented drawl of movie bad guys, said, "Be careful with the fossil, but be even more careful with the box. This is Soviet box. They don't make them like this anymore." Grinning with mischief, he then pulled out a small bottle of dark-colored liquid. "And now we toast with Dagestan cognac," he proclaimed, pouring two glasses, then another two, and then a third round. We toasted to his tyrannosaurs.

Like Brown's first fossil of *Tyrannosaurus rex*, Sasha's dinosaur *Kileskus* was only a fraction of a skeleton. There was part of the snout and the side of the face, a tooth, a chunk of the lower jaw, and some random bones from the hand and feet. These bones were all found within a couple of square meters in a quarry that Sasha's team had been working in for many years, in the Krasnoyarsk region of central Siberia. Krasnoyarsk is one of the more than eighty "federal subjects" of Russia, as the post-Soviet constitution calls the equivalents to American states or Canadian provinces. It's no little Delaware, or even Texas, or incredibly, even Alaska. Krasnoyarsk stretches across nearly the entire midsection of Russia, from the Arctic Sea up north to almost touch the border with Mongolia down south. It's a shade below one million square miles in area, much bigger than Alaska and even slightly larger than Greenland. A lot of space, but very few people: the entire population is about the same as Chicago's. In this vast wilderness, Sasha was able to find the world's oldest tyrannosaur. The name he gave it, *Kileskus*, is based on the word "lizard" in a local language that is spoken by only a few thousand people in this isolated part of the world.

The discovery didn't get much buzz in the press, and it escaped the attention of many scientists when Sasha described it in an obscure Russian journal that isn't on the radar of most paleontologists. *Kileskus* didn't get a funny nickname, and it surely won't be appearing in any future *Jurassic Park* films. It's one of those fifty-some new dinosaurs that are announced in a technical scientific paper every year and then mostly forgotten about, except by a handful of specialist paleontologists. To me, though, *Kileskus* is one of the most interesting discoveries of the last decade, because it is clear proof that tyrannosaurs had gotten an early evolutionary start. *Kileskus* was found in rocks formed during the middle part of the Jurassic Period, about 170 million years ago, more than 100 million years before *T. rex* and its colossal cousins were at the top of their game in North America and Asia.

Kileskus may be important, but it is underwhelming to behold. I first scrutinized the bones in Sasha's dark office, in a grand old building along the icy Neva River, which was still thawing in early April. Yes, Sasha's fossil is only a few bones, but that's not too surprising. The vast majority of new dinosaur discoveries are just a few jumbled pieces of bone, because it takes a whole lot of luck for even a tiny fraction of a skeleton to withstand millions of years buried in the ground. No, what struck me about *Kileskus* was how small it is. All of the bones can comfortably fit into a couple of shoeboxes. I could easily lift them up off the shelf. If I wanted to pick up the skull of *T. rex* back in New York, I would need a forklift.

It's hard to believe that a meek creature like *Kileskus* could have given rise to a giant like *T. rex*. Although an accurate measure of its size is difficult because of its scrappy bones, *Kileskus* was probably only seven or eight feet long, most of that being the skinny tail. It stood a couple of feet tall at most—it would have come up to your waist or chest like a big dog. And it wouldn't have weighed more than a hundred pounds or so. If the forty-foot-long, ten-foot-tall, seven-ton *T. rex* was living in Russia during the Middle Jurassic, it could have brushed *Kileskus* aside with little effort, even with its pathetic little arms. *Kileskus* was not a brutish monster. It wasn't a top predator. It was probably something like a wolf or jackal, a long-legged, lightweight hunter that used speed to chase down small prey. It's surely no coincidence that the quarry in Krasnoyarsk where *Kileskus* was found is bursting with the fossils of small lizards, salamanders, turtles, and mammals. It was these things that the very first tyrannosaurs were eating, not long-necked sauropods or jeep-size stegosaurs.

Because *Kileskus* is so different from *T. rex* in size and hunting habits, how do we know that it's even a tyrannosaur? If *Kileskus* had been discovered at the same time as *T. rex*, scientists probably wouldn't have made the connection. Even if *Kileskus* had been found a few decades ago, it likely wouldn't have registered as a primitive tyrannosaur, a great-great-great-grandparent of *T. rex*. Now we know, and once again, it's because of new fossils.

Sasha had the great fortune of finding *Kileskus* just four years after a team in far western China, led by my colleague Xu Xing, came across a very similar small meat-eater from the middle part of the Jurassic. Thankfully, Xu's team didn't just find a couple of broken bones. They uncovered two nearly complete skeletons, one an adult and the other a teenager. The story of how these dinosaurs got there could be written into the script of a disaster movie. The teenager was found at the bottom of a pit several feet deep, trampled by the adult. They were both engulfed in mud and volcanic ash. Something terrible had clearly happened, but what was torture for these dinosaurs was a lucky break for paleontologists.

Xu and his group named their new dinosaur *Guanlong*, meaning "crown dragon" in Chinese. The name refers to the gaudy Mohawk-like crest of bone that runs along the top of the skull. The crest is thinner than a dinner plate and pierced by a number of holes. It's the type of absurdly impractical-looking thing that probably had only one function: a display ornament for attracting mates and intimidating rivals, kind of like the flamboyant tail of a male peacock, which is for nothing but show.

I spent days poring over the bones of *Guanlong* in Beijing. The crest is what grabbed my attention first, but other features of the bones offer critical clues for placing *Guanlong* on the family tree and linking it to both *Kileskus* and *T. rex*. For a start, it is clearly very similar to *Kileskus*: both are about the same size, have huge windowlike nostrils at the front of the snout, and have long upper-jaw bones with a deep depression above the teeth that would have housed a huge sinus. On the other hand, *Guanlong* exhibits many characteristics that are only seen in *T. rex* and other big tyrannosaurs among all of the meat-eating dinosaurs. In other words, evolutionary novelties, which as we learned earlier, are the key to understanding genealogy. For example, it has heavily fused nasal bones at the top of the snout, a broad and rounded front of the snout, a small horn in front of the eye, and two massive muscle attachment scars on the front of the pelvis. There are many more similarities as well, anatomical minutiae that may seem boring but tell my scientific colleagues and me that *Guanlong* is definitively a primitive tyrannosaur. And because the complete skeletons of *Guanlong* share so many features with the much scrappier bones of *Kileskus*, the latter must be a primitive tyrannosaur as well.

Along with helping to prove that *Kileskus* is a tyrannosaur, the complete skeletons of *Guanlong* also paint a clearer picture of what these earliest and most primitive tyrannosaurs would have looked like, how they behaved, and how they fit into their ecosystems. Based on its limb dimensions—which are known to correlate closely with body weight in living animals—*Guanlong* would have weighed about 70 kilograms, or roughly

150 pounds. *Guanlong* was lithe and lean, with long skinny legs and a tail that stretched far beyond its body for balance. No doubt it was a speedy hunter. It had a mouth full of steak-knife-like teeth befitting a predator, but it also had fairly long arms with three claw-capped fingers capable of grabbing prey with extreme strength. They are totally different from the withered two-fingered arms of *T. rex*.

Guanlong could hunt with its arsenal of speed, sharp teeth, and deadly claws, but it was not a top predator. It lived alongside much larger carnivores like *Monolophosaurus*, which was over fifteen feet long, and *Sinraptor*, a thirty-foot-long close cousin of *Allosaurus* that weighed more than a ton. *Guanlong* lived in the shade of these animals, and probably in fear of them too. At best, *Guanlong* was a second- or third-tier predator, an inconspicuous link in a food chain dominated by other dinosaurs. This would have been the same for *Kileskus* and for some of the other small and primitive tyrannosaurs that have been found recently, like the tiniest one of all, the greyhound-size *Dilong* from China, and *Proceratosaurus*, a dinosaur discovered over a century ago in England that was only recently recognized as an archaic tyrannosaur because it has a Mohawk-like crest similar to *Guanlong*.

These petite tyrannosaurs may not have been much to look at and wouldn't have haunted anyone's nightmares, but they obviously were doing something right. The more fossils we find, the more successful we realize they were. There were a bunch of them, spread all over the world during the approximately 50 million years from the middle part of the Jurassic period well into the Cretaceous, from about 170 until 120 million years ago. They clearly survived that cocktail of environmental and climate changes that felled *Allosaurus*, the sauropods, and the stegosaurs

The skeleton of the dog-size primitive tyrannosaur *Dilong*.

The skull of the human-size primitive tyrannosaur *Guanlong*, showing the gaudy crest of bone on top of its head.

around the Jurassic-Cretaceous boundary. We now have their fossils from throughout Asia, multiple sites in England, the western United States, and probably even Australia. They were able to disperse so widely because they lived when the supercontinent Pangea was still breaking apart, meaning they could easily hop across land bridges that linked the continents, which had yet to move very far away from one another. These early tyrannosaurs had carved out a niche as small to midsize predators living in the underbrush, and they were good at it.

AT SOME POINT, however, tyrannosaurs changed from bit players to the celebrated apex predators that we all love. The first whispers of this transformation are seen in fossils from the early part of the Cretaceous, about 125 million years ago. Most tyrannosaurs living at this time were small. The pint-size *Dilong* is the most extreme example, barely registering on the scales at about twenty pounds. Some were a bit larger, like *Eotyrannus* from England and a few of its older cousins like *Juratyrant* and *Stokesosaurus*, which were bulkier than *Dilong*, *Guanlong*, and *Kileskus*, and maybe reached lengths of about ten to twelve feet and weights of a thousand pounds or so. If you were around back then, and these midsize tyrannosaurs cooperated, you could have ridden them like horses, but they still weren't top-of-the-food-chain animals.

Then in 2009, another piece of the puzzle: a team of Chinese scientists described a highly unusual fossil from the northeastern corner of the country, which they called *Sinotyrannus*. As is so often the case, the new dinosaur was fragmentary: only a small collection of bones was preserved, including the front of the snout and lower jaw, some portions of the backbone, and a few pieces of the hand and pelvis. These bones were really similar to *Guanlong*, and also to *Kileskus*, which would be described a few months later. The base of a tall bony crest was visible right where the snout region was broken, the nostril opening was huge, and there was a deep sinus depression above the teeth. But there was one major difference: *Sinotyrannus* was substantially bigger than *Guanlong*. Based on comparisons to the bones of other meat-eating dinosaurs, it was estimated that this new predator would have been around thirty feet long, and perhaps over a ton in weight. That's the equivalent to at least ten *Guanlongs*. At about 125 million years in age, *Sinotyrannus* was the oldest example of a large-bodied tyrannosaur ever found.

I read the announcement of the new species as a graduate student, about a year after I began my PhD project on carnivorous dinosaur evolution. It was clear to me that the new dinosaur was a tyrannosaur and that it was big, but I didn't know what else to make of it. The fossils were too scrappy to be certain of how large it was or to place it accurately in the family tree. Was it a very close relative of *T. rex*, the first member of that group of really big, deep-skulled, tiny-armed carnivores—*Tyrannosaurus*, *Tarbosaurus*, *Albertosaurus*, *Gorgosaurus*—that dominated the very end of the Cretaceous, about 84 to 66 million years ago? If so, maybe it would tell us how these dinosaur icons became so huge, so dominant. But was it something else? Maybe it was merely a primitive tyrannosaur that outgrew its contemporaries. After all, *Sinotyrannus* lived about 60 million years before *T. rex*, a time when every other tyrannosaur we knew of could fit in the back of a pickup truck.

Could this one find really rewrite tyrannosaur history? I

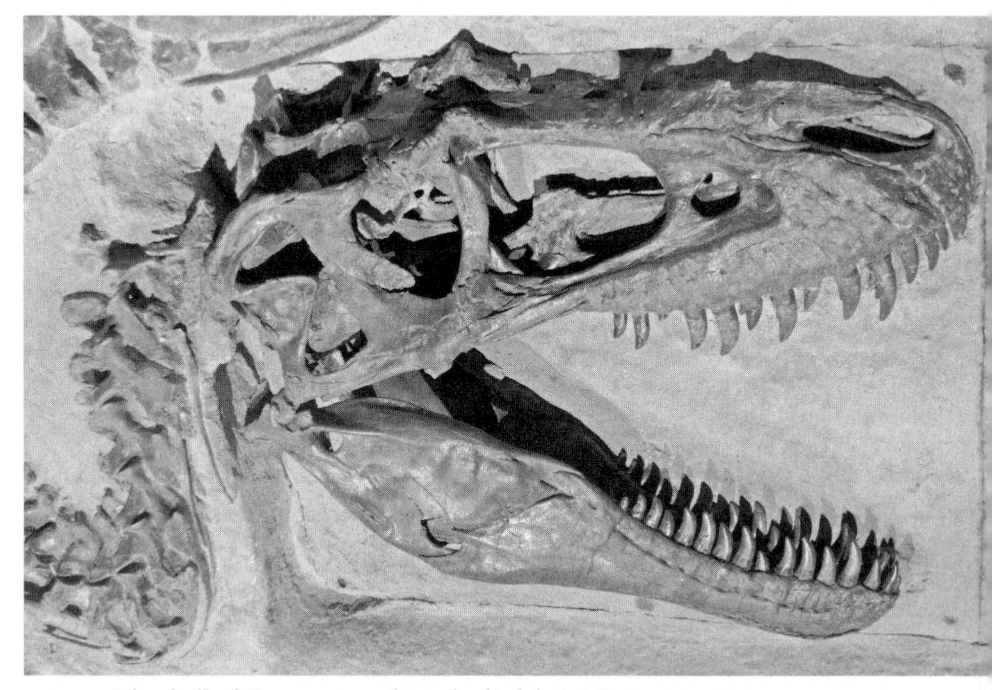

The skull of *Gorgosaurus*, a large-bodied, latest Cretaceous tyrannosaur closely related to *T. rex*.

had the sinking feeling that this fossil would remain a problem for a long time. This happens all too often in the field of dinosaur research: a single fossil emerges that hints at a major evolutionary story—the oldest member of a major group, or the first fossil to exhibit a really important behavior or feature of the skeleton—but it's too broken or incomplete or poorly dated to be certain. Then another fossil is never found and it's just left hanging, a cold case waiting to be solved.

But I shouldn't have been so pessimistic. Just three years later, Xu Xing in China—the man who described *Guanlong* and *Dilong*—published a sensational article in the journal *Nature*. Xu and his team announced yet another new dinosaur, which they called *Yutyrannus*. They had more than just a few bones at their disposal—they had skeletons, three of them. Their new dinosaur was obviously a tyrannosaur and was very similar to *Sinotyrannus*. There were similarities in size and also in the bones—*Yutyrannus* had a flashy head crest and huge nostrils, just like *Sinotyrannus*. *Yutyrannus* was big: the largest skeleton was about thirty feet long. This wasn't an estimate, because Xu and his team could take out a tape measure and size up their new dinosaur, rather than using mathematical equations to guesstimate the size of a complete skeleton based on just a few broken bones, as was our only recourse with *Sinotyrannus*. So *Yutyrannus* sealed the deal: there really were large tyrannosaurs in the Early Cretaceous, at least in China.

There was something else peculiar about *Yutyrannus*. The skeletons were so well preserved that details of the soft tissue were visible. Usually the skin, muscles, and organs decay away long before a fossil is entombed in stone, leaving only the hard parts like bones, teeth, and shells. With *Yutyrannus* we got lucky—these skeletons were buried so quickly, after a volcanic eruption, that some of their softer parts did not decay. Packed all around the bones were dense clusters of slender filaments, each about fifteen centimeters (six inches) long. Similar structures were preserved on the much smaller *Dilong*, which was found in the same rock unit in northeastern China.

These are feathers. Not the quill-pen feathers that make up the wings of today's birds but simpler ones that look more like strands of hair. These were the ancestral structures that bird feathers evolved from, and it is now known that many (and perhaps all) dinosaurs had them. *Yutyrannus* and *Dilong* establish beyond a doubt that tyrannosaurs were among these feathered dinosaurs. Unlike birds, tyrannosaurs certainly were not flying. Instead, they probably used their feathers for display or to keep warm. And because both a large tyrannosaur like *Yutyrannus* and a small tyrannosaur like *Dilong* have feathers, this implies that the common ancestor of all tyrannosaurs had feathers, and therefore that the great *T. rex* itself was most likely feathered, too.

The fluff-covered skeletons of *Yutyrannus* launched this new dinosaur to stardom in the international press, but feathers are a story that we'll come back to later. For me, the real importance of *Yutyrannus* was that it could help us better understand how tyrannosaurs evolved into their huge sizes. *Yutyrannus* and *Sinotyrannus* were big—much larger than any other tyrannosaurs living before the very end of the Cretaceous, when *T. rex* and its brethren reigned supreme. However, these two Chinese tyrannosaurs weren't truly colossal: they were about the same size as *Allosaurus* or the big predator *Sinraptor* that preyed on *Guanlong*, nowhere near the monstrous forty-foot-long, seven-ton body sizes of *T. rex* and its very close relatives. Not only that, but when the complete skeletons of *Yutyrannus* are compared bone by bone with the skeletons of *T. rex*, it becomes clear they are quite different. *Yutyrannus* looks like an overgrown version of *Guanlong*, with its ornamental head crest, big nostrils, and long, three-fingered hands. It doesn't have the deep muscular skull, thick railroad-spike teeth, and pathetic arms of *T. rex*.

This leads to an unexpected conclusion: despite their big bodies, *Yutyrannus* and *Sinotyrannus* weren't very closely related to *T. rex*, and they didn't have much to do with the evolution of colossal sizes in the latest Cretaceous tyrannosaurs. Instead, they were primitive tyrannosaurs experimenting with large body sizes independent of their later cousins. Put another way, they were evolutionary dead ends that, as far as we know, didn't exist outside of one corner of China during the early Cretaceous. (This assertion can of course be proven wrong with new discoveries.) They lived alongside small tyrannosaurs, which were by far the more common type thriving in the Jurassic and early Cretaceous times.

Even though they were not directly ancestral to *T. rex*, *Yutyrannus* and *Sinotyrannus* are far from unimportant. These early Cretaceous species do show that tyrannosaurs had the capability to become big fairly early in their evolution. *Yutyrannus* and *Sinotyrannus* were, as far as we know, the largest predators in their ecosystems. They were at the top of the food chain, the lords of a lush forest—humid in the summer, liable to be buried by snow in the winter—that clung to the sides of steep volcanoes, alive with the chirps of primitive birds and raptor dinosaurs with feathers. They had their choice of prey: corpulent long-necked sauropods if they were feeling particularly hungry, or a bounty of sheep-size, beaked plant-eaters called *Psittacosaurus*, primitive cousins of *Triceratops*, which 60 million years later would battle *T. rex* itself on the floodplains of western North America.

In other places, separated in time and space from the forests of Early Cretaceous China, where the species of tyrannosaurs were small or medium-size, they were dwarfed by larger predators. *Sinraptor* towered over *Guanlong* in the Middle Jurassic of China. *Allosaurus* outmuscled the mule-size tyrannosaur *Stokesosaurus* in the later Jurassic of North America. The carcharodontosaur *Neovenator* held down *Eotyrannus* in the Early Cretaceous of England. And there are many more examples. It seems tyrannosaurs could get big if they had the opportunity, but only if there were no larger predators around.

THE QUESTION REMAINS: how did *T. rex* and its closest relatives shoot up to such mind-boggling sizes? We need to look into the fossil record to see when the very first truly huge tyrannosaurs with the classic *T. rex* body plan emerged. By this, I mean tyrannosaurs that were over thirty-five feet long and one and a half tons in weight, with the big deep skulls, muscular jaws, banana-size teeth, pathetic arms, and bulky leg muscles that define *T. rex*.

This type of tyrannosaur—true giants, undoubted top predators of record size—made their first appearance in western North America about 84 to 80 million years ago. Once they began to appear, they started turning up everywhere, both in North America and Asia. Clearly an explosive diversification had occurred.

We know that the big switch happened some time in the middle part of the Cretaceous, between about 110 and 84 million years ago. Before this time, there were many small to midsize tyrannosaurs living all over the world, with only a few random bigger species like *Yutyrannus*. After this time, enormous tyrannosaurs reigned throughout North America and Asia, but only those continents, and no species smaller than a minibus remained. This was a dramatic change, one of the biggest in the entire history of dinosaurs. Frustratingly, very few fossils record what was going on. The middle Cretaceous is something of a dark period in dinosaur evolution. By pure bad luck, very few fossils from this entire 25-million-year time span have been found. So we're left scratching our heads, like a detective tasked with solving a crime when the crime scene preserves no fingerprints, DNA data, or tangible evidence of any kind.

What we can say, based on our growing understanding of

what the Earth was like during the middle Cretaceous, is that this was probably not a great time to be a dinosaur. About 94 million years ago, between the Cenomanian and Turonian subdivisions of the Cretaceous Period, there was a spasm of environmental change. Temperatures spiked, sea levels violently oscillated, and the deep oceans were starved of oxygen. We don't yet know why this happened, but one of the leading theories is that a surge of volcanic activity belched enormous quantities of carbon dioxide and other noxious gases into the atmosphere, causing a runaway greenhouse effect and poisoning the planet. Whatever their causes, these environmental changes triggered a mass extinction. It wasn't as big as the extinctions at the ends of the Permian and Triassic periods, which helped dinosaurs rise to dominance, but something more akin to what happened across the Jurassic-Cretaceous boundary. Still, it was one of the worst mass die-offs during the Age of Dinosaurs. Many ocean-living invertebrates disappeared for good, as did various types of reptiles.

The extremely poor middle Cretaceous fossil record has made it difficult to know how these environmental dramas affected dinosaurs. However, paleontologists have recently managed to pry important new specimens from this gap. A pattern is clearly coming into focus: none of the large predators from this 25-million-year time window are tyrannosaurs. All of them belong to other groups of big carnivores like the ceratosaurs, spinosaurs, and especially the carcharodontosaurs. This latter group of ultrapredators, which (as we saw in the previous chapter) utterly dominated the Early Cretaceous, continued their reign deep into the middle Cretaceous. The thirty-five-foot-long carcharodontosaur *Siats* was the top predator in western North America about 98.5 million years ago. In Asia, the nearly *T. rex*–size *Chilantaisaurus* and the smaller *Shaochilong* were the top guns about 92 million years ago, and in South America, carcharodontosaurs like *Aerosteon* reigned about 85 million years ago.

The tyrannosaurs that lived alongside these carcharodontosaurs, on the other hand, still weren't very special, at least in their outward appearance. We don't have many of their fossils, but some have started to turn up recently. The best of them come from Uzbekistan, where Sasha Averianov and his colleague Hans-Dieter Sues—a German-born paleontologist with an ever-present smile and infectious laugh, who is now a senior researcher at the Smithsonian Institution—worked for over a decade, in the barren Kyzylkum Desert.

That Soviet-era box that Sasha carefully handed over to me a few years ago contained some of these bones. The reason I took them back to Edinburgh to CAT-scan was because two of these specimens were braincases—the puzzle of fused bones at the back of the skull that surrounded the brain and ear. If you want to see inside these braincases, into the cavities that housed the brain and sense organs, you could cut open the braincase with a saw, which is what Osborn did with the first *T. rex* skull, damaging it forever in the name of science. Nowadays we can use the CAT scanner and its high-powered X-rays, and we don't have to damage a thing. When we scanned the Uzbek braincases, we confirmed that they belonged to a tyrannosaur, as they had the same architecture of bones surrounding the spinal cord and the same long tube-shaped brain cavity of *T. rex*, *Albertosaurus*, and other tyrannosaurs. They even had a middle ear with a very long cochlea, another signature tyrannosaur feature, which allowed these predators to better hear low-frequency sounds. However, the Uzbek tyrannosaur was still a Mini-Me, just about the size of a horse.

In spring 2016, Sasha, Hans, and I gave the Uzbek tyrannosaur a formal scientific name, *Timurlengia euotica*. The name honors Timur, also known as Tamerlane, the infamous Central Asian warlord who ruled over Uzbekistan and many of the surrounding lands in the fourteenth century. It's a fitting name for a tyrannosaur, even a midsize one that was still a few rungs below the top of the food ladder. Although not a colossus, *Timurlengia* was developing a larger brain and more sophisticated senses—heightened smell, vision, and hearing—than other meat-eating dinosaurs, adaptations that would eventually turn out to be handy predatory weapons for the huge tyrannosaurs that came later. Tyrannosaurs were becoming smart before they got big, but no matter how clever they were, *Timurlengia* and its comrades were still living under the thumb of the real warlords of the middle Cretaceous, the carcharodontosaurs.

Then, when the clock struck 84 million years ago and the fossil record became rich again, the carcharodontosaurs were gone in North America and Asia, replaced by monstrous tyrannosaurs. A major evolutionary turnover had occurred. Was this due to the lingering effects of the temperature and sea-level changes that occurred at the Cenomanian-Turonian boundary? Was it sudden or gradual? Did tyrannosaurs actively outcompete the carcharodontosaurs, muscling them into extinction or outsmarting them with their big brains and keenly developed senses? Or did environmental changes cause these other large predators to go extinct but spare tyrannosaurs, which then opportunistically took over the large predator role? We just don't have enough evidence to know for certain, but whatever the answer, there is no denying that by the dawn of the Campanian subinterval of the latest Cretaceous, beginning about 84 million years ago, tyrannosaurs had risen to the top of the food pyramid.

During the final 20 million years of the Cretaceous, tyrannosaurs flourished, ruling the river valleys, lakeshores, floodplains, forests, and deserts of North America and Asia. There is no mistaking their signature look: huge head, athletic body, sad arms, muscular legs, long tail. They bit so hard that they crunched through the bones of their prey; they grew so fast that they put on about five pounds every day during their teenage years; and they lived so hard that we have yet to find an individual that was more than thirty years old when it died. And they were impressively diverse: we have found nearly twenty species of these big-boned tyrannosaurs from the latest Cretaceous, and there are surely many more out there waiting to be discovered. The Pinocchio-nosed *Qianzhousaurus*, so fortuitously discovered by that still-anonymous backhoe operator at the Chinese construction site, is one of the latest examples. Just as Brown and Osborn grasped over a hundred years ago, when they were the first humans to set eyes on a tyrannosaur, *T. rex* and its brethren really were the kings of the dinosaur world.

The world they lorded over was very different from the planet in which tyrannosaurs grew up. Back when *Kileskus*, *Guanlong*, and *Yutyrannus* were stalking prey, the supercontinent Pangea had only recently begun to split, so tyrannosaurs could migrate easily across the Earth. By the latest Cretaceous, however, the continents had drifted much farther apart, reaching positions similar to the ones they occupy today. A map from this time would have looked quite a bit like today's globe. There were, however, some major differences. Due to sea-level rise in the Late Cretaceous, North America was bisected by a seaway stretching from the Arctic to the Gulf of Mexico, and a flooded Europe was reduced to a smattering of small islands. *T. rex*'s Earth was a fragmented planet, with different groups of dinosaurs living in separate areas. As a result, champions in one region might not be able to conquer another for one simple reason: they couldn't get there. Colossal tyrannosaurs never seemed to gain a foothold in Europe or the southern continents, where other groups of large predators prospered, but in North America and Asia, tyrannosaurs were unrivaled. They had become the transcendent terrors that fire our imaginations.

The **KING** *of the* **DINOSAURS**

Tyrannosaurus rex

THE *TRICERATOPS* WAS SAFE. It was across the river, separated by impassable rapids from the danger brewing on the opposite bank. But it could see what was about to happen and was powerless to stop it.

No more than fifty feet away, on a spike of sand and mud that jutted into the other side of the water, a group of three *Edmontosaurus*es lingered. Their sharp ducklike bills plucked leaves from the flowery shrubs clinging to the shore. Their cheeks—heavy with nourishment—rocked side to side in a chewing motion. The late evening sun shimmered across the currents, and the whistles of birds high in the trees radiated peace and calm.

But all was not OK. On the far shore, the *Triceratops* noticed something the *Edmontosaurus* herd could not—another creature, hiding in the taller trees at the edge of the jungle where it met the sand bar, its green scaly skin almost perfectly camouflaged. Its eyes gave it away: two bulbous spheres, sparkling with anticipation. They darted from left to right, in split-second intervals, surveying the three unaware plant-munchers. Waiting for the right moment.

And then it came, in a burst of violence.

The red-eyed, green-skinned monster pounced out of the brush and into the path of the plant-eaters. It was a terrifying sight: the lurking predator was longer than a city bus. It reached forty feet (thirteen meters) long and weighed at least five tons. Fluff stuck out of the scales of its neck and back—a mangy, hairy fuzz. Its tail was long and muscular, its legs stocky, its arms laughably tiny, dangling to the side as it lunged toward the *Edmontosaurus* pack headfirst, jaws agape.

When it opened its mouth, there were about fifty pointy teeth inside, each the size of a railroad spike. They clamped down on the tail of one of the *Edmontosaurus*es, the cacophony of crunching bone and shrieking anguish echoing through the forest.

Desperate, the assaulted *Edmontosaurus* wrestled itself free and waddled off into the trees, its severed tail dangling behind it, carrying a broken tooth from the predator as a battle scar. Would it survive or succumb to its injuries in the hidden depths of the forest? The *Triceratops* would never know.

Annoyed by its failed attack, the beast turned its attention to the smallest of the duckbills, but the youth was racing away into the woods, dodging trunks and bushes at a sprinter's speed. The bulky carnivore realized it had no hope of catching it and emitted a deep-throated wail in frustration.

There was still one *Edmontosaurus* left, cornered on the sandbar: water on one side, the meat-lusting monster on the other. As the predator turned its head back toward the river, the two of them locked eyes. Escape was impossible, and then the inevitable happened.

The head darted forward. Teeth met flesh. Bones shattered as the neck of the herbivore was ripped apart, blood spilling into the water and mixing with the white foam currents, the broken teeth of the predator raining through the sky as it tore at its victim.

Then, from back in the forest, there was a rustling noise. Branches snapped and leaves flew about. The *Triceratops* watched in awe as four other big-headed, spike-toothed green brutes—nearly identical in size and shape to the first one—bounded onto the riverbank. They were a pack; the attacker was their leader, and now the underlings got to share in its victory. The five hungry creatures snorted and snarled, nipping at each other and biting each other's faces as they jockeyed for the best cuts of meat.

From the comfort of the opposite shore, the *Triceratops* knew exactly what it was seeing. For it had been there before—it had once escaped the jaws of one of these voracious killers, goring it with one of its horns until the beast released its grip. This feared predator was known to all *Triceratops*. It was their great rival, the terror that would rush like a ghost from the trees and mow down entire herds. It was *Tyrannosaurus rex*—the King of the Dinosaurs, the largest predator that has ever lived on land in the 4.5-billion-year history of Earth.

T. REX IS a celebrity character—the nightmare haunter—but it was also a real animal. Paleontologists know quite a lot about it: what it looked like, how it moved and breathed and sensed its world, what it ate, how it grew, and why it was able to get so big. In part, that's because we have a lot of fossils: over fifty skeletons, some nearly complete, more than for almost any other dinosaur. But more than anything, it's because so many scientists are impulsively drawn to the majesty that is the King, the way so many people are obsessed with movie stars and athletes. When scientists get infatuated with something, we start playing around with every instrument, experiment, or other type of analysis at our disposal. We've thrown the whole toolbox at *T. rex*: CAT scans to look into its brain and sense organs, computer animations to understand its posture and locomotion, engineering software to model how it ate, microscopic study of its bones to see how it grew, and the list goes on. As a result, we know more about this Cretaceous dinosaur than we do about many living animals.

What was *T. rex* like as a living, breathing, feeding, moving, growing animal? Let me indulge you with an unauthorized biography of the King of Dinosaurs.

Let's start with the vital stats.

It goes without saying, but *T. rex* was huge: adults were about forty-two feet (thirteen meters) long and weighed in the ballpark of seven or eight tons, based on those equations from a few chapters ago, which calculate body weight from the thickness of the thighbone. These proportions are off the charts for carnivorous dinosaurs. The rulers of the Jurassic—the Butcher *Allosaurus*, *Torvosaurus*, and their kin—got up to about thirty-three feet (ten meters) long and a few tons—monsters to be sure, but they had nothing on Rex. After temperature and sea-level changes ushered in the Cretaceous, some of the carcharodontosaurs from Africa and South America got even bigger than their Jurassic predecessors. *Giganotosaurus*, for example, was about as long as *T. rex* and may have reached about six tons. But that's still a good ton or two lighter than Rex, so the King stands alone as the biggest purely meat-eating animal that lived on land during the time of dinosaurs, or indeed at any time in the history of our planet.

Tyrannosaurus rex skeleton at the American Museum of Natural History in New York.

Show a picture of *T. rex* to kindergartners and they'll immediately know what it is. It has a signature style, a unique physique, or in scientific parlance, a distinctive body plan. The head was enormous, perched on a neck short and stout like a bodybuilder's. Balancing the oversize noggin was a long, tapering tail that stuck out horizontally like a seesaw. Rex stood only on its hind legs, its muscular thighs and calves powering its movements. Like a ballerina, it balanced on the tips of its feet, the arch or sole rarely touching the ground, all of its weight held by its massive three toes. The forelimbs looked useless: puny things with two stubby fingers, comically out of proportion to the rest of the body. And the body itself: not fat like one of the long-necked sauropods, but not the skinny frame of a fast-running *Velociraptor* either. Its very own body type.

The seat of Rex's power was its head. It was a killing machine, a torture chamber for its prey, and an evil mask all in one. At around five feet long from snout to ear, the skull was nearly the length of an average person. More than fifty knife-sharp teeth made for a sinister smile. There were little nipping teeth at the front of the snout and a row of serrated spikes the size and shape of bananas running along the sides of the upper and lower jaws. Muscles to open and close those jaws bulged out of the back of the head near the bottle-cap-size hole that served as the ear. Each eyeball was the size of a grapefruit. In front of it, but covered in skin, was a massive sinus system that helped to lighten the head, and then big fleshy horns at the tip of the snout. Small horns protruded in front of and behind each eye, and another stuck downwards from each cheek—gnarly knobs of bone covered in keratin, the same stuff that makes up our fingernails. Imagine this hideous visage as your last memory before the teeth came crushing down, breaking your bones. Many a dinosaur met its end that way.

Covering the body—the head, the wee arms, the stocky legs, all the way to tip of the tail—was a thick, scaly hide. In this way, *T. rex* resembled an overgrown crocodile or an iguana—lizardlike. But there was one key difference: Rex also had feathers sticking out from between its scales. As mentioned in the last chapter, these were not big branching ones like those on a bird wing, but were simpler filaments that looked and felt more like hair, the larger ones stiff like the quills of a porcupine. *T. rex* certainly couldn't fly, and neither did its ancestors that first evolved these proto-feathers, way back in the early days of the dinosaurs. No, as we'll learn later, feathers started out as simple wisps of integument, which creatures like *T. rex* used to keep warm, and as displays to attract mates and scare off rivals. Paleontologists have yet to find any fossilized feathers on a *T. rex* skeleton, but we're confident that it must have had some fluff because primitive tyrannosaurs—*Dilong* and *Yutyrannus*, which we met last chapter—have been found coated in hairlike feathers, as have many other types of theropods preserved in those rare conditions that allow soft bits to turn into fossils. That means that the ancestors of *T. rex* had feathers, so it is highly likely that Rex did too.

T. rex lived from about 68 to 66 million years ago, and its dominion was the forest-covered coastal plains and river valleys of western North America. There it lorded over diverse ecosystems that included a bounty of prey species: the horn-faced *Triceratops*, the duck-billed *Edmontosaurus*, the tanklike *Ankylosaurus*, the dome-headed *Pachycephalosaurus*, and many more. Its only competition for food was from the much smaller dromaeosaurs—raptor dinosaurs à la *Velociraptor*—which is to say it didn't have much competition at all.

Although several other tyrannosaurs had thrived in these same environments during the preceding 10 to 15 million years, they were not the ancestors of *T. rex*. Instead, Rex's closest cousins were Asian species like *Tarbosaurus* and *Zhuchengtyrannus*. *T. rex*, as it turns out, was an immigrant. It got its start in China or Mongolia, hopped across the Bering Land Bridge, journeyed through Alaska and Canada, and made its way down into the heart of what's now America. When the young Rex arrived at its new home, it found things ripe for the taking. It swept across western North America, an invasive pest that spread all the way from Canada down to New Mexico and Texas, elbowing out all of the other midsize to large predatory dinosaurs so that it alone controlled an entire continent.

Then one day it all ended. *T. rex* was there when the asteroid fell down from the sky 66 million years ago, putting a violent end to the Cretaceous, exterminating all of the nonflying dinosaurs. That's a story we'll get to later. For the time being, only one fact really matters: the King went out on top, cut down at the peak of its power.

WHAT FEAST BEFITS the King? We know *T. rex* was a carnivore of the highest order, a pure meat-eater. It's one of the simplest inferences that we can make about any dinosaur, and it doesn't require any fancy experiments or machines to figure out. *T. rex* had a mouth full of thick, serrated, razor-sharp teeth. Its hands and feet boasted big pointy claws. There's really only one reason an animal would have these things: they're weapons, used to procure and process flesh. If your teeth look like knives and your fingers and toes are hooks, then you're not eating cabbages. For anybody who doubts that, there is plenty of other evidence: bones have been found preserved in the stomach area of tyrannosaur skeletons and in the coprolites (fossilized dung) dropped by tyrannosaurs, and western North America is peppered with skeletons of plant-eating dinosaurs—particularly *Triceratops* and *Edmontosaurus*—with bite marks that perfectly match the size and shape of *T. rex* teeth.

Like so many monarchs, Rex was a glutton. It devoured meat. Scientists have predicted how much food an adult *T. rex* would need to survive, based on the food intake of living predators scaled up to an animal of Rex's size. The estimates are nauseating. If *T. rex* had the metabolism of a reptile, then it would have required about 12 pounds (5.5 kilograms) of *Triceratops* chops per day. But that's very likely a vast underestimate, because as we'll see later, dinosaurs were much more birdlike than reptilian in their behaviors and physiology, and they (or at least many of them) may have even been warm-blooded like us. If that was the case, then Rex needed to gobble up some 250 pounds (about 111 kilograms) of grub each and every day. That's many tens of thousands of calories, maybe even hundreds of thousands, depending on how fatty the King liked its steak. It's roughly the same amount of food eaten by three or four large male lions, some of the most energetic, and hungriest, modern carnivores.

Maybe you've heard the rumor that *T. rex* liked its meat dead and rotten, that Rex was a scavenger, a seven-ton carcass collector too slow, too stupid, or too big to hunt for its own fresh food. This accusation seems to make the rounds every few years, one of those stories that science reporters can't get enough of. Don't believe it. It defies common sense that an agile and energetic animal with a knife-toothed head nearly the size of a Smart car wouldn't use its well-endowed anatomy to take down prey but would just walk around picking up leftovers. It also runs against what we know about modern carnivores: very few meat-eaters are pure scavengers, and the outliers that do it well—vultures, for instance—are fliers that can survey wide areas from above and swoop down whenever they see (or smell) a decaying body. Most carnivores, on the other hand, actively hunt but also scavenge whenever they have the chance. After all, who turns down a free meal? That's true of lions, leopards, wolves, even hyenas, which are not the pure scavengers of legend but actually earn much of their food through the chase. Like these animals, *T. rex* was probably both a hunter and an opportunistic scavenger.

Still doubt that Rex went out and got its own food? There's

fossil evidence that proves *T. rex* hunted, at least part of the time. Many of those *Triceratops* and *Edmontosaurus* bones pockmarked with *T. rex* tooth impressions show signs of healing and regrowth, so they must have been attacked while alive but survived. The most provocative of these specimens is a set of two fused *Edmontosaurus* tailbones with a *T. rex* tooth stuck between them, enveloped by the gnarly mass of scar tissue that merged the two bones together as they healed. The poor duck-billed dinosaur was viciously attacked by a tyrannosaur and left with a terrible injury, but it kept the predator's tooth as a trophy from its near-death experience.

Many of Rex's bite marks are peculiar. Most theropods left simple feeding traces on the bones of their prey: long, parallel, shallow scratches, a sign that the teeth were just barely kissing the bone. That's not surprising, because even though dinosaurs could replace their teeth throughout life (unlike us), no predator would want to break its chompers every time it ate. *T. rex* was different, though. Its bite marks are more complex: they start with a deep circular puncture, like a bullet hole, which grades into an elongate furrow. This is a sign that Rex bit deeply into its victim, often right through the bones, and then ripped back. Paleontologists have come up with a special term for this style of eating: puncture-pull feeding. During the puncture phase of its bite, Rex clamped down hard enough to literally break through the bones of its prey. This is why the fossilized dung heaps left by *T. rex* are chock full of bony chunks. Bone crunching is not normal—some mammals, like hyenas, do it, but most modern reptiles do not. As far as we know, big tyrannosaurs like *T. rex* were the only dinosaurs capable of it. It was one of the powers that made the King an ultimate killing machine.

How was *T. rex* able to do it? For starters, its teeth were perfectly adapted. The thick, peglike teeth were strong enough that they wouldn't easily break when they hit bone. Next, consider the power behind those teeth: *T. rex*'s jaw muscles were massive, bulging mounds of sinew that provided enough energy to shatter the limbs, backs, and necks of *Triceratops*es, *Edmontosaurus*es, and other prey. We can tell that Rex had some of the largest and most powerful jaw muscles of any dinosaur, based on the very broad and deep gullies on the skull bones where the muscles attached.

Experiments can simulate the actions of these jaw muscles. One of my colleagues, Greg Erickson of Florida State University, designed a particularly clever experiment in the mid-1990s, right after he finished graduate school. Greg is one of my favorite people to hang around with—he talks with the cadence of high school jock and often looks the part in his worn baseball cap, cold beer in hand. A few years back, Greg was a regular talking head on a cable TV program about weird animal incidents—alligators crawling through sewers and invading trailer parks, that kind of thing. As much fun as he is, I admire Greg deeply as a scientist, because he brings a different approach to paleontology—experimental, quantitative, rigorously grounded in comparisons to modern animals.

Greg spends a lot of time with engineers, and one day they came up with a crazy idea: they would rig up a laboratory version of *T. rex* and determine how strong its bite was. They started with a *Triceratops* pelvis with a half-inch-deep puncture left by a Rex, and then asked a simple question: how much force would it take to make an indentation this deep? They couldn't take a real *T. rex* and make it bite into a real *Triceratops*, but they found a way to simulate it by making a bronze and aluminum cast of a *T. rex* tooth, putting it into a hydraulic loading machine, and smashing it into the pelvis of a cow, which is very similar in shape and structure to the *Triceratops* bone. They pushed and pushed the tooth until it made a half-inch-deep hole, and then used their instruments to read out how much force it required: 13,400 newtons, equivalent to about 3,000 pounds.

That's a staggering number—about the weight of an old-school pickup truck. By comparison, humans exert a maximum force of about 175 pounds with our rear teeth, and African lions bite at about 940 pounds. The only modern land animals that come close to *T. rex* are alligators, which also bite at around 3,000 pounds. However, we need to remember that the 3,000-pound figure for *T. rex* is for only a single tooth—imagine how much power a mouth full of these railroad spikes would have delivered! And because it's a measure of the force required to make one observed fossil bite mark, it's likely that this is an underestimate of the maximal biting power. Rex probably had the strongest bite of any land animal that ever lived. It could crunch bones with ease and would have been strong enough to bite through a car.

All of that strength came from the jaw muscles; they were the engine that powered the teeth to deliver the bone-breaking bite. But that's not the entire story. If the muscles delivered enough force to bust the bones of prey, they could have also broken the skull bones of the *T. rex* itself. Basic physics: every action has an equal and opposite reaction. So it wasn't enough for *T. rex* to have massive teeth and huge jaw muscles—it also needed a skull that could withstand the tremendous stresses that occurred each time it snapped its jaws shut.

To figure out how, we needed to turn back to the engineers and to another paleontologist who has crossed over into the realm of hard-core numbers science. Emily Rayfield's lab at the University of Bristol in England is a big bright room with a row of computers, its large windows and breezy open plan like something out of Silicon Valley. The shelves are lined with manuals for various software packages, but there's nary a fossil in sight. Emily doesn't often collect fossils; she's not that kind of paleontologist. Instead, she builds computer models of fossils—say, the skull of *T. rex*—and uses a technique called finite element analysis (FEA) to study how they would have behaved in a mechanical sense.

FEA was developed by engineers and calculates the stress and strain distributions in a digital model of a structure when it is subjected to various simulated loads. In plain English, it's a way to predict what will happen to something when some kind of force is applied to it. This is very useful for engineers. Before a construction crew starts building a bridge, let's say, the engineers better be pretty damn sure that the bridge isn't going to collapse when heavy cars start driving over it. To check, they can build a digital model of the bridge and use the computer to imitate the stresses from real cars to see how the bridge reacts. Does it absorb the weight and force of the cars easily, or does it start to crack under pressure? If it does start to crack, the computer can identify the weak points and the engineers can go back to the plans for the actual bridge to make the necessary fixes.

Emily does the same thing with dinosaurs, and *T. rex* has been one of her favorite muses. She built a digital model of Rex's skull based on CAT scans of a well-preserved fossil, and then used the FEA program to simulate the forces of a bone-

The skull of *Tyrannosaurus rex*. *Courtesy of Larry Witmer.*

crunching bite and analyze how the skull reacted. The verdict: *T. rex* had a remarkably strong skull that was optimized to endure the extreme pushing and pulling forces of its three-thousand-pound-per-tooth chomp. It was built like an airplane fuselage: the individual bones tightly sutured together so that they wouldn't come apart when the stress hit. The nasal bones above the snout were fused together into a long, vaulted tube, which acted as a stress sink. Thick bars of bone around the eye provided strength and rigidity, and the robust lower jaw was almost circular in cross section so that it could withstand high

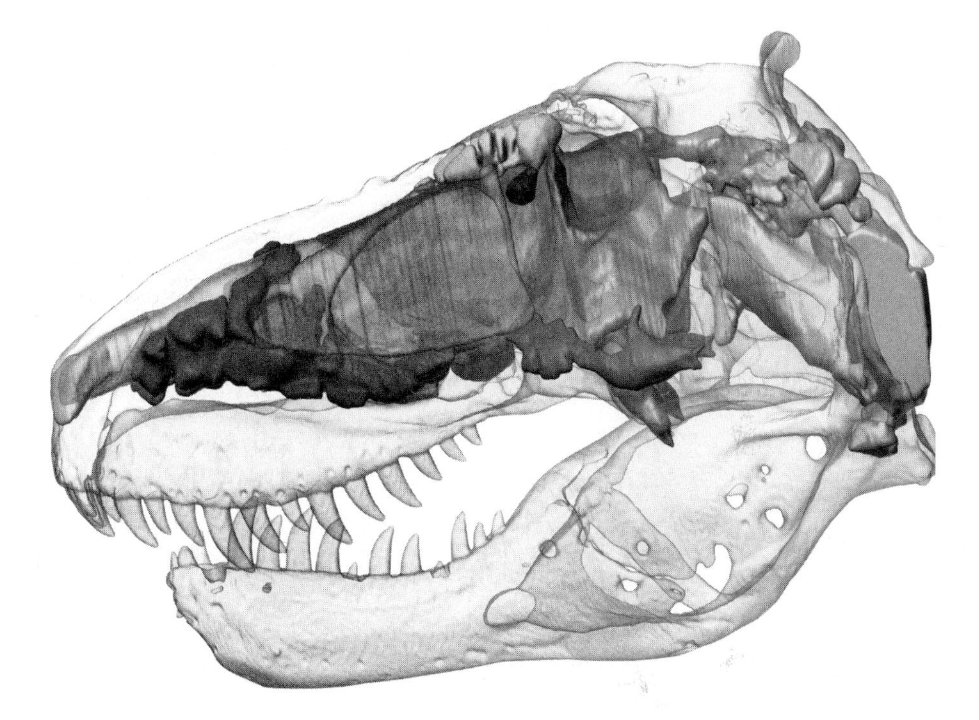

The brain cavity (upper right-hand corner) and sinuses inside the skull of a *Tyrannosaurus rex*, revealed by CAT scans. *Courtesy of Larry Witmer.*

pressures from all directions. None of these things are present in other theropods, which had daintier skulls with looser connections among the various bones.

That's the final piece of the puzzle, the last component in the tool kit that allowed *T. rex* to bite so strongly that it punctured, and then pulled through, the bones of its supper. Thick peg-like teeth, huge jaw muscles, and a rigidly constructed skull: that was the winning combination. Without any of these things, *T. rex* would have been a normal theropod, slicing and dicing its prey with care. That's how the other big boys did it—*Allosaurus*,

Torvosaurus, and the carcharodontosaurs—because they didn't have the arsenal necessary for bone-crunching. Once again, the King stands alone.

T. REX WAS able to gnash through most anything that it wanted to eat, whether it was splurging on a forty-foot-long *Edmontosaurus* or snacking on smaller contemporaries like the donkey-size ornithischian *Thescelosaurus*. But how did it capture its food?

Not, as it turns out, with exceptional speed. *T. rex* was a special dinosaur in many ways, but one thing it could not do is move very fast. There's a famous scene in *Jurassic Park* where the bloodthirsty *T. rex*, convulsed by its insatiable appetite for human flesh, chases down a jeep driving at highway speeds. Don't believe the movie magic—the real *T. rex* likely would have been left in the dust once the jeep got up to third gear. It's not that Rex was a plodding slouch that waddled through the forest. Far from it—*T. rex* was agile and energetic, and it moved with purpose, its head and tail balancing each other as it tiptoed through the trees, stalking its prey. But its maximum speed was probably in the ballpark of ten to twenty-five miles per hour. That's faster than we can run, but it's not as quick as a racehorse or, certainly, a car on the open road.

Once again, it's high-tech computer modeling that has allowed paleontologists to study how *T. rex* moved. This work was pioneered in the early 2000s by John Hutchinson, an American transplant to England who is now a professor at the Royal Veterinary College near London. He spends his days working with animals: monitoring the livestock on his university's research campus, making elephants run across scales to study their posture and locomotion, dissecting ostriches and giraffes and other exotic creatures. John chronicles his adventures on his popular blog, the wonderfully but somewhat disturbingly titled *What's in John's Freezer?* He also pops up frequently as a talking head on television documentaries, often adorned in his favorite purple shirt, which somehow doesn't crack the cameras with its glare. Like Greg Erickson, John is a scientist whom I've long admired because of his unique angle on studying dinosaurs. For John, the present is very much the key to the past: find out as much as we can about the anatomy and behaviors of today's animals, and that will help us understand dinosaurs.

If you visit John's lab, he really does have freezers stocked with the frozen cadavers of animals of all shapes and sizes, from all over the world. Odds are, one or two of them will be out thawing, getting ready for the dissection table. But there is a more sterile side to John's lab: the computers, which he uses to make digital models of dinosaurs, like those we saw in chapter 3 that we made to predict the weight and posture of long-necked sauropods. He starts with a three-dimensional model of a skeleton, captured through CAT scans, laser surface scans, or the photogrammetry method we learned about earlier. Then he uses his knowledge of modern animals to flesh it out: to add muscles (whose sizes and positions are based on the attachment sites visible on the fossil bones) and other soft tissues, wrap it up in skin, and position it in realistic postures. The computer does its magic, putting the model through all sorts of gymnastics routines, and calculates how fast the real animal was likely able to move. John's modeling provides us with the range of ten to twenty-five miles per hour that I cited for *T. rex*'s speed.

The computer models also make clear that Rex would have needed absurdly large leg muscles to run as fast as a horse: more than 85 percent of its total body mass in its thighs alone, which is obviously impossible. *T. rex* was simply too big to run exceptionally fast. Its sheer size also conferred another liability: the Tyrant King couldn't turn very quickly, or otherwise it would topple over like a truck taking a corner too sharply. Thus, the reality is, Spielberg had it wrong, *T. rex* was no sprinter, and it would have ambushed its prey with a quick strike rather than chasing it down like a cheetah.

Ambushing prey can take a lot of energy—in bursts. Thankfully, *T. rex* had another trick up its sleeve, or more precisely, deep inside its chest. Remember those hyperefficient lungs of sauropods, which allowed them to reach such enormous sizes? *T. rex* had the same lungs. They are the lungs of today's birds: rigid bellows anchored to the backbone, able to extract oxygen when the animal breaths in *and* also when it breathes out. They're different from our lungs, which can take in oxygen only during inhalation, then spew out carbon dioxide during exhalation. They are a stunning feat of biological engineering. When today's birds—and also *T. rex*—breathe in, oxygen-rich air courses through the lungs as you would expect. However, some of the inhaled air doesn't go through the lungs right away but is shunted into a system of sacs connected to the lung. There it waits, until it is released when the animal exhales, passing through the lungs and delivering its oxygen-rich hit even as carbon dioxide waste is being expelled. Birds get twice the bang for the buck, a continuous supply of energy-sustaining oxygen. If you've ever wondered how some birds can fly at tens of thousands of feet, in rarefied air where we would have a hard time breathing (just ask anyone who has experienced the oxygen masks coming down midflight), their lungs are their secret weapon.

Paleontologists have yet to find a fossilized *T. rex* lung and probably never will. The thin tissues are too delicate to fossilize. But we know that Rex had a birdlike, ultra-efficient lung, because this kind of breathing system leaves impressions on the bones, which do fossilize. It all has to do with the air sacs, the air-storage compartments integral to the bird-style lung. These sacs are akin to balloons: they are soft, thin-walled, compliant bags that inflate and deflate during the ventilation cycle. Many air sacs are connected to the lung, nestled in between the many other organs of the chest, including the trachea and esophagus, the heart, the stomach, and intestines. Sometimes they run out of room and start wiggling their way into the only space still available: the bones themselves. As they do so, they invade the bone through large, smooth-walled holes and then expand into chambers once inside. These signatures are easy to identify on fossils. We see them on the backbones of *T. rex*, along with many other dinosaurs, including, as we learned about earlier, the humongous sauropods. We never see these things on mammals, or lizards, or frogs, or fish, or any other types of animals—only in modern birds and extinct dinosaurs and a few very close relatives, a telltale fingerprint of their unique lungs.

The drama of a *T. rex* ambush is coming into focus. The lungs delivered the energy, which was then transferred to the leg muscles, which propelled the Rex forward with a burst of speed to lunge at its startled victim. And then what happened? Just imagine *T. rex* as a giant land shark. Like a Great White, all of the action was with its head. Rex led with its noggin and used its clamp-strong jaws to grab its dinner, subdue it, kill it, and crunch through its flesh and guts and bones before swallowing. *T. rex* simply had to hunt headfirst, because its arms were pitifully tiny. The King evolved from smaller tyrannosaur ancestors, like *Guanlong* and *Dilong*, that used their much longer arms to grab their prey. But during the course of tyrannosaur evolution, the head got bigger, the arms got smaller, and the skull gradually took over all of the hunting functions that the arms used to perform.

Why, then, did *T. rex* still have arms? Why didn't it lose them completely, the way whales ditched their no-longer-necessary hindlegs when they evolved from land mammals that colonized the water? That mystery has captivated scientists for a long time, and it's kept cartoonists and comedians supplied with an endless source of material for bad puns. As it turns out, those little arms—as silly as they may look—were not useless. Although short, they were stocky and muscular, and they served a purpose.

Sara Burch figured it out. Sara and I both trained in Paul Sereno's lab at the University of Chicago, where we became friends, but our paths diverged afterward: I went down the route of studying genealogy and evolution, and Sara became enthralled with bones and muscles. She did her PhD in an anatomy department, where she dissected a zoo's worth of animals, and has since carved out a career that is common for paleontologists: teaching human anatomy to medical students. Sara knows more about the anatomical structure of dinosaurs than almost anyone alive—how their bones connected to each other, what kind of muscles they had. She reconstructed the forearm muscles of *T. rex* and many other theropods, determining which muscles were present, and how big they were, from the preserved attachment sites on the bones, helped along by comparisons to modern reptiles and birds for guidance. Rex's seemingly sad arms actually turned out to have powerful shoulder extensors and elbow flexors—exactly those muscles needed to hold on to something that is trying to pull away, to keep it close to the chest. It seems that *T. rex* used its short but strong arms to hold down struggling prey while the jaws did their bone-crunching thing. The arms were accessories to murder.

Now there's one final twist in the story of how *T. rex* hunted. We increasingly believe that Rex didn't go on the prowl alone; it traveled in packs. The evidence comes from a Canadian fossil site located between Edmonton and Calgary, in what is now Dry Island Buffalo Jump Provincial Park. It was discovered back in 1910 by none other than Barnum Brown, who just a few years earlier found the first *T. rex* skeleton in Montana. Brown was traveling through the heart of the Canadian prairies, floating down the Red Deer River on a boat and dropping anchor wherever he saw dinosaur bones sticking out of the riverbank. When he came to Dry Island, he noticed a number of bones from a slightly older cousin of *T. rex* called *Albertosaurus*, one of the North American apex predators right before Rex migrated over from Asia. He had time to collect only a small sample before heading back to New York.

Those bones languished deep in the vaults of the American Museum for decades, until Phil Currie—Canada's leading dinosaur hunter (and one of the nicest guys you'll ever meet)—took notice of them in the 1990s. He retraced Brown's steps, relocated the site, and began excavating. Over the next decade, his team collected more than a thousand bones, which belong to at least a dozen individuals, ranging from youngsters to adults, all of them *Albertosaurus*. There's really only one way numerous individuals of the same species can be preserved together: they must have lived and died together. A few years later, Phil's crew found a similar mass graveyard in Mongolia, packed with several *Tarbosaurus*es, the very closest Asian cousin of *T. rex*. *Albertosaurus* and *Tarbosaurus* were evidently pack animals, and we reckon that Rex itself was as well. If a seven-ton, bone-crunching, ambush predator isn't scary enough on its own, then just imagine a pack of them working together. Sweet dreams!

LET'S GET INTO the King's head. What did it think? How did it sense its world? How did it locate its prey? These are, of course, very difficult questions to answer. Even with modern living animals, it's almost impossible to put ourselves in their feet, or paws, or paddles and feel what their world is like. But we can study their brains and sense organs and start to put together a picture. With dinosaurs, however, we are usually out of luck: the brains, eyes, nerves, and tissues associated with the ears and nose are soft and decay easily, meaning they rarely make it through the rigors of fossilization. What can we do?

Technology, yet again, makes the impossible possible. The brains, ears, noses, and eyes of dinosaurs may be long gone, but these organs occupied spaces in the bones. The brain cavity, the eye socket, and so on. We can study these spaces to get a sense of the original sense organs that filled them, but there is another problem: many of these spaces are inside the bones, not observable from the outside. That's where the technology comes in: we can use CAT scans (also known by the shorter abbreviation of CT) to visualize the inside of dinosaur bones. CAT scans are nothing more than high-powered X-rays. That's why they're

Ian Butler CAT-scanning the skull of the primitive tyrannosaur *Timurlengia* at the University of Edinburgh.

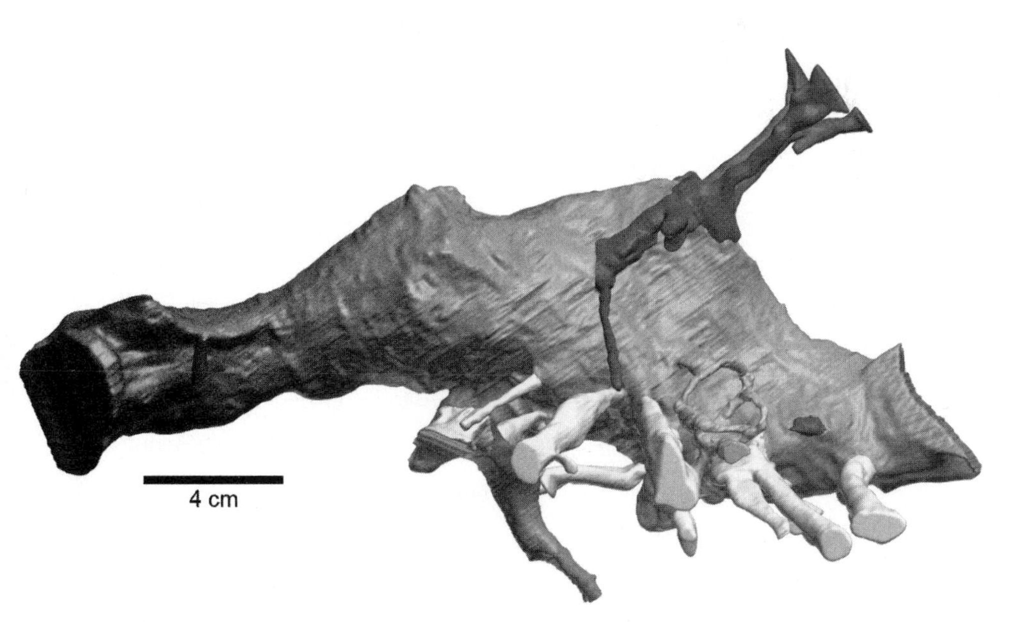

A CAT scan reconstruction of the brain, inner ear, and associated nerves and blood vessels of *Tyrannosaurus rex*. *Courtesy of Larry Witmer.*

popular in medicine: if you feel a pain in your gut or a creak in your bones, your doctor will probably stick you in a CAT scanner to see what's going on inside your body without having to cut you open. Ditto with dinosaurs. We can use the X-rays to take an array of internal images, which we can then stitch together into three-dimensional models using various software packages. This procedure has become practically routine in paleontology, such that many labs—including my own in Edinburgh—have a CAT scanner onsite. Ours was hand-built by one of my colleagues, Ian Butler, a geochemist by training who now finds himself scanning fossil after fossil, each one leading him deeper into the addiction that is paleontology.

Ian and I are newcomers to the fossil-scanning game. We're following in the footsteps of a few giants in the field: Larry Witmer of Ohio University, Chris Brochu of the University of Iowa, and the wife-and-husband team of Amy Balanoff and Gabe Bever, who started at the University of Texas, moved on to the American Museum in New York (where I met them when I was a PhD student), and are now ensconced at Johns Hopkins University in Baltimore. Balanoff and Bever are virtuosos who can read CAT scans the way a linguist deciphers ancient manuscripts. In the grayscale splotches of the X-rays, they can make out the internal structures that powered the intelligence and sensory prowess of long-dead dinosaurs. Tyrannosaurs like *T. rex* have been some of their favorite subjects—their favorite patients, if you will, whose behaviors and cognitive abilities are mysteries to be diagnosed.

The scans tell us quite a bit about our patient. First off, Rex had a distinctive brain. It didn't look anything like our brain but was more of a long tube with a slight kink at its back, surrounded by an extensive network of sinuses. It's also a relatively large brain, at least for a dinosaur, which hints that *T. rex* was fairly intelligent. Now, measuring intelligence is riddled with uncertainties, even for humans: just think of all of the IQ tests, exams, SAT scores, and other things that we use to try to assess how smart people are. However, there is a straightforward measure that scientists use to roughly compare the intelligence of different animals. It's called the encephalization quotient (EQ). It's basically a measure of the relative size of the brain compared to the size of the body (because, after all, bigger animals have bigger brains simply because of their body size: elephants have bigger brains than we do but are not more intelligent). The largest tyrannosaurs like *T. rex* had an EQ in the range of 2.0 to 2.4. By comparison, our EQ is about 7.5, dolphins come in around 4.0 to 4.5, chimps at about 2.2 to 2.5, dogs and cats are in the 1.0 to 1.2 range, and mice and rats languish around 0.5. Based on these numbers, we can say that Rex was roughly as smart as a chimp and more intelligent than dogs and cats. That's a whole lot smarter than the dinosaurs of stereotype.

One part of the tyrannosaur brain was particularly enlarged: the olfactory bulbs. These are the lobes at the front of the brain that control the sense of smell. The two bulbs were each a little larger than a golf ball, much bigger in absolute size than in any other theropods. Of course, *T. rex* was one of the biggest theropods, so maybe it had whopping olfactory lobes simply by virtue of its extreme bulk. What is needed, then, is a relative measure of olfactory bulb size. My friend Darla Zelenitsky of the University of Calgary did just that. She compiled CAT scans of numerous theropods, calculated the size of their olfactory bulbs, and normalized them by dividing by body size. Even after all of this, she still found the big tyrannosaurs to be extreme outliers: they, along with the raptor dinosaurs, had proportionally enormous olfactory bulbs, and thus a sharp sense of smell, compared to other meat-eating dinosaurs.

It wasn't only the nose. Other senses were heightened as well. The CAT scans allow us to see inside Rex's inner ear: the pretzel-shaped network of tubes that control both hearing and balance. The semicircular canals at the top of the inner ear—which make the pretzel shape—were long and loopy. As we know from comparisons to modern animals, this means that *T. rex* was agile and capable of highly coordinated head and eye movements. Sticking downward from the pretzel is the cochlea, the part of the inner ear that regulates hearing. In *T. rex* the cochlea was elongated, more than in most other dinosaurs. There is a tight relationship in living animals: the longer the cochlea, the better sensitivity to lower-frequency sounds. In other words, Rex also had a keen sense of hearing. Vision, too: the huge eyeballs of *T. rex* faced partially to the side and partially to the front, meaning that they were capable of binocular vision. The King could see in three dimensions and perceive depth, just like us. There's another scene in *Jurassic Park* where the freaked-out humans are told to stay still, because if they don't move, then the *T. rex* can't see them. Nonsense—because it could sense depth, a real Rex would have made an easy meal out of those sad, misinformed people.

Thus it wasn't all brute strength. *T. rex* had brawn all right, but it also had brains. High intelligence, world-class sense of smell, keen hearing and vision. Add these things to the armory: they're what Rex used to target its victims, to choose which poor dinosaurs would have to die.

WHEN I ENVISION *T. rex* as a real animal, what most amazes me is that it would have started life as a tiny hatchling. All dinosaurs, as far as we know, hatched from eggs. We have yet to find any *T. rex* eggs, but we do have eggs and nests of many closely related theropods. Most of these dinosaurs seemed to guard their nests and provide at least a bit of care for their young. Without some parental love, the baby dinosaurs would have been hopeless, because they were tiny: no dinosaur eggs that we know of are larger than a basketball, so even the mightiest species like *T. rex* would have been, at most, the size of a pigeon when they entered the world.

Back when my parents were learning about dinosaurs in school, the assumption was that *T. rex* and kin grew like iguanas: they kept growing throughout their life, gradually getting bigger and bigger and bigger. Rex was able to get so large because it lived for a long time: after about a century, it would reach its final size of forty-two feet and seven tons, then finally saunter off and die. This type of thinking even percolated into the dinosaur books I read as a child, but like many once cherished notions about dinosaurs, it turns out to be false. Dinosaurs like *T. rex* grew rapidly, a lot more like birds than lizards.

The evidence is buried deep inside the bones of dinosaurs, and paleontologists like Greg Erickson found a way to tease it out. Bones are not static rods and blobs stuck in our bodies; no, they're dynamic, growing, living tissues that repair and remodel themselves constantly. This is why your bones heal if you break them. As most bones grow, they get wider in all directions, expanding outward from the center, but usually bones grow rapidly only during certain parts of the year: the summer or the wet season, when food is plentiful. Growth slows down

Tyrannosaurus rex skeleton on display at the Royal Tyrrell Museum in Alberta, Canada.

during the winter or dry season. If you cut open a bone, you can see a record of each time growth transitions from rapid to slow: a ring. That's right—just like trees, bones have rings inside, and because that summer-to-winter switch happens once a year, that means one ring is laid down each year. By counting the rings you can tell how old a dinosaur was when it died.

Greg got permission to cut open the bones of several different *T. rex* skeletons, along with many other close tyrannosaur relatives like *Albertosaurus* and *Gorgosaurus*. Shockingly, not a single bone had more than thirty growth rings. That means tyrannosaurs matured, reached adult size, and died within three decades. Big dinosaurs like *T. rex* didn't grow slowly for many decades (or centuries) but must have reached their huge sizes by growing rapidly for a much shorter period of time. But how quickly? To figure it out, Greg constructed growth curves: he plotted the age of each skeleton, determined from the number of bone rings against its body size, calculated from those equations we learned about earlier that estimate weight based on limb dimensions. This allowed Greg to compute how quickly *T. rex* grew each year. The number is almost too big to comprehend: during its teenage years, from about ages ten to twenty, Rex put on about 1,700 pounds (760 kilograms) per year. That's close to 5 pounds per day! No wonder *T. rex* had to eat so much—all of that *Edmontosaurus* and *Triceratops* flesh fired the insane teenage growth spurt that turned a kitty-size hatchling into the King of the Dinosaurs.

You could call *T. rex* the James Dean of dinosaurs: it lived fast and died young. And all of that hard living put a tremendous strain on its body. The skeleton had to endure the daily addition of five pounds during the spurt years. Somehow the body had to morph from wee hatchling to monster, so it comes as no surprise that the skeleton of *T. rex* changed dramatically as it matured. As youngsters, they were sleek cheetahs, as teenagers gangly looking sprinters, and as adults pure-blooded terrors longer and heavier than a bus. The younger ones probably ran a lot faster than the adults and maybe could have chased down their prey, whereas the silverbacks were so huge that they could only ambush and relied much more on their strength than their speed. What's particularly frightening is that juveniles and adults seemed to live together in packs, meaning they may have hunted in teams, complementing each other's skills to make life hell on their prey.

One of my dearest paleontologist friends has made a career studying how *T. rex* changed as it grew. He's a Canadian named Thomas Carr, now a professor at Wisconsin's Carthage College. You can spot Thomas from a mile away. He has the fashion sense of a 1970s preacher and some of the mannerisms of Sheldon Cooper from *The Big Bang Theory*. Thomas always wears black velvet suits, usually with a black or dark red shirt underneath. He has long bushy sideburns and a mop of light hair. A silver skull ring adorns his hand. He's easily consumed by things and has a long-running obsession with absinthe and the Doors. That and tyrannosaurs: he'll talk a lot about *T. rex*, because it's his favorite subject of all. Ever since he was young, he wanted to study the Tyrant King, and he eventually wrote a PhD dissertation on how the skull of *T. rex* changed as it matured. It was over 1,270 pages long; meticulous as Thomas always is, it's one of his shorter scholarly works.

Bone by bone, Thomas has chronicled the metamorphosis of *Tyrannosaurus rex*. Almost the entire head was reshaped as it went from boy to man, girl to woman. The skull started out long and low, with a stretched-out snout, thin teeth, and shallow depressions for jaw muscles. Throughout the teenage years, it got bigger, deeper, and stronger. The sutures between bones locked more tightly together, the jaw-muscle depressions became much deeper, and the teeth turned into bone-shattering pegs. The juveniles weren't capable of puncture-pull feeding; that only became possible in adulthood, around the same time that Rex switched from a speedster to a slower ambusher. There were other changes too: the sinuses within the skull expanded, probably to help lighten the ever heavier head, and the little horns on the eyes and cheeks became larger and more prominent, the tiny bumps becoming gaudy display ornaments to attract mates when those teenage hormones kicked in.

It was quite the transformation. After all of those meals, the decade of exponential growth, the complete refiguring of the skull, the loss of the ability to run fast but the acquisition of puncture-pull biting, the Rex was all man, all woman, and ready to claim its throne.

AND THERE YOU have it, a glimpse into the life and times of the most famous dinosaur in history. *T. rex* bit so hard it could crunch through the bones of its prey, it was so bulky that it couldn't run fast as an adult, it grew so fast as a teenager that it put on five pounds a day for a decade, it had a big brain and sharp senses, it hung around in packs, and it was even covered in feathers. Maybe it's not the biography you were expecting. And there's the rub. Everything we have learned about *T. rex* tells us that it, and dinosaurs more generally, were incredible feats of evolution, well adapted to their environments, the rulers of their time. Far from being failures, they were evolutionary success stories. They were also remarkably similar to animals of today, particularly birds—Rex had feathers, grew rapidly, and even breathed like a bird. Dinosaurs were not alien creatures. No, they were real animals that had to do what all animals do: grow, eat, move, and reproduce. And none of them did it better than *T. rex*, the one true King.

7

DINOSAURS *at the* TOP *of* THEIR GAME

Triceratops

AS TERRIFYING AS IT WAS, *T. rex* was not a global supervillain. Its dominion was North America—western North America, to be more precise. No Asian, European, or South American dinosaurs lived in fear of *T. rex*. In fact, they never would have met one.

During the latest Cretaceous—the last throes of dinosaur evolution, about 84 to 66 million years ago, when *T. rex* and its jumbo-size tyrannosaur cousins topped the food chain—the geographical harmony of Pangea was a distant memory. By then, the supercontinent had long ago fractured into pieces, each chunk drifting apart from the others slowly over the Jurassic and Early to middle part of the Cretaceous, the gaps in between the new shards of land filled by oceans. When *T. rex* took its crown, just a couple million years before the Age of Dinosaurs ended in a bang, the map was more or less as it is today.

North of the equator there were two big landmasses: North America and Asia, with essentially their modern shapes. They ever so slightly kissed each other near the North Pole, but otherwise were separated by a wide Pacific Ocean. There was an Atlantic Ocean, too, on the other side of North America, which encircled a series of islands that corresponded to modern-day Europe. Sea level was so high during the latest Cretaceous—the result of a hothouse world where very little, if any, water was locked up in polar ice caps—that most of low-lying Europe was flooded. Only a constellation of random morsels—the higher parts of Europe—poked up from the waves. High sea level also pushed water farther inland, so that warm subtropical seas lapped far onto both North America and Asia. The North American seaway extended all the way from the Gulf of Mexico to the Arctic. In effect, it bisected the continent into an eastern slice called Appalachia and a western microcontinent called Laramidia, the hunting grounds of *T. rex*.

It was a similar situation in the south. The yin-and-yang puzzle pieces of South America and Africa had just recently detached, a narrow corridor of the South Atlantic nestled between. Antarctica sat at the bottom of the world, balanced on the South Pole. Off to its north was Australia, a bit more crescent-shaped than it is today. Fingers of crust kept Antarctica in contact with both Australia and South America, but these were tenuous, liable to be swamped any time sea level crept up slightly. During those high-water stands, just as in the north, seas extended far inland onto the southern continents, drowning much of northern Africa and southern South America. What is now the Sahara would have been waterlogged. However, during those times when the seas receded a little bit, an archipelago provided a route between Africa and Europe—a highway, albeit a fleeting and treacherous one, between north and south.

A few hundred miles off the east coast of Africa was a triangular wedge, an island continent. This was India, the only large piece of land in the latest Cretaceous that would look out of place to us today. India began its life as a sliver of ancient Gondwana—the big mass of southern lands that separated from the north when Pangea began to split—wedged in between what would become Africa and Antarctica. It severed all ties with its neighbors some time during the early part of the Cretaceous and began a race northward, moving at more than six inches (fifteen centimeters) per year. Most continents, by contrast, drift at a much slower pace, about the speed that our fingernails grow. This brought India to the middle part of the proto–Indian Ocean, a bit south of the Horn of Africa, in the latest Cretaceous. Another 10 million years or so and it would complete its journey, colliding with Asia to form the Himalayas, but by then the dinosaurs were long gone.

In between these pieces of land were the oceans—a domain dinosaurs were never able to conquer. The warm waters of the Cretaceous, as during the Jurassic and Triassic beforehand, were the hunting grounds of various types of giant reptiles: plesiosaurs with long noodle-shaped necks, pliosaurs with enormous heads and paddlelike flippers, streamlined and finned creatures called ichthyosaurs that looked like reptilian versions of dolphins, and many others. They dined on each other and on fish and sharks (most of which were much smaller than today's species), which in turn fed on tiny shelled plankton that choked the ocean currents. None of these reptiles were dinosaurs—even though they are often mistaken for dinosaurs in popular books and movies, they were merely distant reptilian cousins. For whatever reason—and we don't yet know the answer—no dinosaurs were able to do what whales did: start on the land, change their bodies into swimming machines, and make a living in the water.

They were stuck on the land, one of the few liabilities they were never able to overcome. In the latest Cretaceous, this meant that they had to deal with a disjointed world. The land was divided into different kingdoms, fragments of dry ground separated by those reptile-infested seas, their dinosaurs isolated from each other. And that includes *T. rex*. The King may have been able to easily subjugate the dinosaurs of Europe or India or South America, but it never got the chance. It was restricted to western North America.

This was good news for other dinosaurs, especially the plant-eaters, but it also gave other types of meat-eaters the opportunity to seize their own kingdoms, and various groups of carnivores did just that, the story a little bit different on each of the Cretaceous continents. Each landmass had a unique suite of dinosaurs—its own megapredators, second-tier hunters, scavengers, big and small herbivores, and omnivores. Provinciality extended to other species as well: there were distinct types of crocodiles, turtles, lizards, frogs, and fishes on the various parcels of land, and of course, different types of plants too. In this way, isolation bred diversification.

So it was that the latest Cretaceous—this world of such geographical and ecological complexity, with different ecosystems stranded on different continents—was the heyday of the dinosaurs. It was their time of greatest diversity, the apogee of their success. There were more species than ever before, from pint-size ones to giants, eating all kinds of foods, endowed with a spectacular variety of crests, horns, spikes, feathers, claws, and teeth. Dinosaurs at the top of their game, doing as well or better than they had ever done, still in control more than 150 million years after their earliest ancestors were born on Pangea.

TO FIND THE best fossils of latest Cretaceous dinosaurs—including bones of *T. rex* itself—you have to go to hell—or rather, the badlands surrounding Hell Creek, a once trickling tributary of the Missouri River that is now a flooded arm of a reservoir in northeastern Montana. It's a place of stifling humidity and mosquito swarms, with rare breezes and little shade. Just rock bluffs that stretch to the horizon in all directions, radiating heat like a sauna.

Barnum Brown was one of the first explorers to visit Hell Creek in search of dinosaurs, and it was in the scabby hills a hundred miles or so southeast of the creek where he found the first skeleton of *T. rex* in 1902. His bosses in New York were overjoyed, and Brown was given a mandate to bring more fossils back to the big city. Over the next few years, decked out in his fur coat with his pickaxe slung over his shoulder, he prospected the bluffs, gullies, and dry streambeds along the Missouri River and farther southeast. The fossils kept coming, and after a while Brown came to understand the geology of the area. All of the bones were buried inside a thick sequence of rocks that formed much of the badlands topography—a layer-cake array of reds, oranges, browns, tans, and blacks, made up of sand and mud deposited by ancient rivers. He called these rocks the Hell Creek Formation.

The Hell Creek rocks were formed between about 67 and 66 million years ago, by a tangle of rivers that drained the young Rocky Mountains to the west, then meandered across a vast floodplain, occasionally bursting their banks and pooling into lakes and swamps, before emptying eastward into that great seaway that cut North America in two. These were fertile, lush environments, a perfect setting for so many types of dinosaurs to thrive. It was also an environment where sediments were being deposited and turned into rock, with the bones inside them along for the ride. Lots of dinosaurs and lots of sediments—that's the recipe for a fossil bonanza.

I took my first trip to Hell in 2005, a century after Brown's *T. rex* was unveiled in New York. I was an undergraduate, a month removed from my first-ever dinosaur-hunting expedition, excavating Jurassic sauropods in Wyoming with Paul

Sereno. Looking to gain additional fieldwork experience, I drove out to Montana with a crew from the closest thing I could call my local museum, the aforementioned Burpee Museum of Natural History in Rockford, Illinois.

Rockford isn't the type of place you'd expect to have a dinosaur museum. For one, not a single dinosaur fossil has ever been found in Illinois—my home state is too flat, too geologically boring, almost barren of rocks formed during the time dinosaurs reigned. Nor have past decades been kind to its manufacturing-based economy. Yet Rockford has one of the finest natural history museums in the Midwest. The staff of the Burpee Museum often refer to themselves as "the little museum that could," which speaks to the odd twists of fate they've had to navigate. For most of its existence, the museum was little more than a fusty collection of stuffed birds, rocks, and Native American arrowheads, poking out of the nooks and lofts of a once-grand nineteenth-century mansion. Then in the 1990s, the museum received a startling donation from a private benefactor, and a new wing was added. Exhibits were needed to fill the expansion, so the administrators hatched a trip to Hell Creek to bring back dinosaurs.

At that time, the Burpee Museum had only a single paleontology curator on its payroll, a soft-spoken, barrel-chested northern Illinois boy named Mike Henderson, infatuated with the smeared fossils of worms that lived hundreds of millions of years before the dinosaurs. He needed help, so he teamed up with a childhood friend—a boisterous, loudmouthed people person named Scott Williams. Along with comic books and superhero movies, Scott loved dinosaurs as a kid, but he didn't have the opportunity to pursue paleontology as a career and ended up going into law enforcement. He was still a cop—and he looked the part, with his goatee, stocky build, and thick Chicago accent—when I first met him at the Burpee Museum when I was in high school. A few years later, after leaving the force for a full-time career in science, he became the collections manager at the museum, and today he helps manage one of the world's largest dinosaur collections, at the Museum of the Rockies in Montana.

During the summer of 2001, Mike and Scott led an eclectic crew of museum staff, geology students, and amateur volunteers out to the heart of Hell. They set up camp near the tiny town of Ekalaka, Montana, population about three hundred, not too far from the T-shaped junction where Montana meets both Dakotas. Brown had once searched this ground, but Mike and Scott found something that had eluded even the maestro. They happened upon the best, most complete skeleton of a teenage *T. rex* that had ever been found. It was *the* keystone fossil that told paleontologists that the King was a gangly, long-snouted, thin-toothed sprinter as a youngster, before it metamorphosed into a truck-size bone-crunching brute as an adult.

Mike, Scott, and their crew discovered a fossil that immediately made the Burpee Museum a major player in dinosaur research. When the skeleton—which they nicknamed Jane, after a museum donor—went on display a few years later, paleontologists from around the world flocked to anonymous Rockford, Illinois, to see it—as did many hundreds of thousands of kids, families, and tourists. The Burpee Museum now had a superstar to headline its new exhibition hall.

Mike and Scott kept going back to Hell for months at a time during the next few summers. Eventually they invited me to come along, but only after I earned their trust. I had become friends with Mike and Scott during my frequent visits to the Burpee Museum, which began while I was a high school sophomore. They first knew me as an annoying teenager with a dinosaur obsession, who, tape recorder and autograph Sharpie in hand, religiously attended the Museum's annual PaleoFest, where notable scientists came to speak about their adventures studying dinosaurs (which, incidentally, is where I first met two of the eminent paleontologists who would later become my academic advisors: Paul Sereno and Mark Norell). I continued to drive up to Rockford throughout college, and once I started formally training to become a paleontologist in Sereno's lab, Mike and Scott thought I was ready to join them on their annual descent into Hell.

A thousand miles separate Rockford and Ekalaka. When we arrived, we took up residence at a place called Camp Needmore, a scattering of wooden bunkhouses deep in the cool pine forests that rise above the badlands. That first night I was kept awake by the wail of a synthesizer, coming from one of the cabins next door. It was the bunkhouse occupied by a trio of volunteers who drove out separately from Rockford, all professionals taking a break from the grind of the office. Their ringleader was a short, quirky fellow. His name—Helmuth Redschlag—conjured up images of an imperious Prussian general, but he was from Middle America, and his job was much more sedate: he was an architect. Each night he partied deep into the morning with his friends—feasting on filet mignon and imported Italian cheeses, sipping fruity Belgian beers to the disco trash beat. Still, every morning he was up at six A.M., eager to head back into the furnace of Hell on the trail of dinosaurs.

"It makes me feel alive. The heat. The sun beaming down, burning you, scarring your neck and your back, desperate for shade and water," Helmuth said to me in the calm of one morning, before we set out into the inferno. Uh huh, uh huh, I nodded along, unsure of what to make of him.

A couple of days later, while I was out prospecting with Scott and some of the other student volunteers, we got a frantic call from Helmuth. He was wandering a few miles down the road, enjoying the pain of the sun on his skin, when something caught his eye in a gully: a dark brown bulge sticking out of the dull tan-colored mud rocks. A lot of things caught Helmuth's eye—he was an architect, after all, and a fine one at that—and his attention to the details of shapes and textures made him a very good fossil hunter. He sensed that this one was special, so he started to dig into the hillside. By the time we arrived on the scene, he had exposed a thighbone, several ribs and vertebrae, and part of the skull of a dinosaur. The bones from the head gave away its identity. Many of them were randomly shaped pieces of something flat and platelike, resembling shattered glass, and a few others were sharp, pointy cones: horns. Only one dinosaur in the Hell Creek ecosystem fit the profile: *Triceratops*, with three horns on its face and a broad, thick, billboard-like frill extending from behind its eyes.

Triceratops, like its arch-nemesis *T. rex*, is a dinosaur icon. In films and documentaries, it usually plays the gentle, sympathetic plant-eater, the perfect dramatic foil to the Tyrant King. Sherlock versus Moriarty, Batman versus the Joker, Trike versus Rex. But it's not all movie magic; no, these two dinosaurs truly would have been rivals 66 million years ago. They lived together along the lakes and rivers of the Hell Creek world, and they were the two most common species there—*Triceratops* making up some 40 percent of Hell Creek dinosaur fossils, *T. rex* coming in second at about 25 percent. The King needed immense amounts of flesh to fuel its metabolism; its three-horned comrade was fourteen tons of slow-moving prime steak. You can figure out what happened next. Indeed, *Triceratops* bones with bite marks matching *T. rex* attest to their ancient battles, but don't think for a moment that it was an unfair fight, always destined to go the way of the predator. *Triceratops* was armed with a set of weapons: its horns, a stout one on the nose and a longer, thinner one above each eye. Like the frill on the back of the head, the horns probably evolved primarily for display—to make *Triceratops* seem sexy to potential mates and scary to its rivals—but no doubt *Triceratops* would use them in self-defense when needed.

Triceratops is a new type of dinosaur in our story. It belongs to a group of plant-eating ornithischians called ceratopsians, which descended from some of the small, fast-running, leaf-toothed critters like *Heterodontosaurus* and *Lesothosaurus* of the Early Jurassic. Beginning some time in the Jurassic, the ceratopsians went down their own evolutionary path. They switched from walking on their hind legs to plodding along on all fours and started to develop a wardrobe variety of horns and frills on their heads, which would get larger and gaudier as a hatchling turned into a hormone-fueled adult that needed to woo mates. The first ceratopsians were dog-size critters; one of them, *Leptoceratops*, straggled into the Late Cretaceous, where it lived alongside *Triceratops*, its much larger cousin. As ceratopsids got bigger over time—morphing into bovine versions of dinosaurs that were very common in North America during the latest Cretaceous—they changed their jaws so that they could engulf unholy quan-

The skull of *Triceratops*, the iconic horned dinosaur.

tities of plants. They packed their teeth closely together so that the jaws were essentially blades—four in all, one on each side of the upper jaw and one on each side of the lower. The jaws would snap shut in a simple up-and-down motion, the opposing blades slicing past each other like a guillotine. At the front of the snout was a razor-sharp beak, which would pluck the stems and leaves and deliver them to the blades. *Triceratops* surely was as good at eating plants as *T. rex* was at devouring meat.

Finding a *Triceratops* was another coup for the Burpee Museum, exactly what it needed to accompany the teenage *T. rex* in the new exhibit space. From the moment Helmuth showed us the bones in the ground, I could tell that Mike and Scott were thinking exactly that. Helmuth too—and as the discoverer of

A jumble of *Triceratops* bones at the Homer site, belonging to a pack of juveniles.

Map of Camp Homer Site

July 22, 2005 Steve Brusatte

Specimen Number	Description	Strike and Dip	Jacket #
BMR 2005 CH 1	? cylindrical w/ flange	10° dip 28° strike	
BMR 2005 CH 2	? rib head	9° dip no definite long axis	
BMR 2005 CH 3	contain in baggie vert w/ neural arch	–	
BMR 2005 CH 4	rib shaft	–	
BMR 2005 CH 5	rib shaft	–	
BMR 2005 CH 6	rib shaft	–	
BMR 2005 CH 7	femur	11° dip 353° strike	
BMR 2005 CH 8	rib	8.5° dip 15° strike	
BMR 2005 CH 9	flat platy bone	20° dip no definite long axis	
BMR 2005 CH 10	flat platy bone	21° dip 47° strike	
BMR 2005 CH 11	flat platy bone	66° dip no def. long axis	
BMR 2005 CH 12	rib shaft	31° dip 27° strike	
BMR 2005 CH 13	flat platy bone	35° dip no def. long axis	
BMR 2005 CH 14	unident.	–	
BMR 2005 CH 15	rib shaft	19° dip 80° strike	
BMR 2005 CH 16	? arm bone	13.5° dip 354° strike	

My field notebook from the 2005 Burpee Museum expedition to Hell Creek, showing the field map of the Homer *Triceratops* site that I made.

the new dinosaur, he got to give it a nickname. Like me he is a big fan of *The Simpsons*, so he decided to call it Homer. One day, we surmised, Homer would join Jane in the halls of the Burpee Museum.

But first we had to get Homer out of the ground. The crew began to wrap up the exposed bones in plaster bandages, to protect them during transport back to Rockford. Others were tasked with finding more bones. Thomas Carr—my absinthe-drinking, Goth-dressing friend who studies *T. rex*—was with us on the expedition and was part of this team. Clad in khaki (it was far too hot for his usual all-black getup) and sucking down Gatorade by the gallon (absinthe was more of an indoor pursuit), he attacked the mudstones with his rock hammer (which he nicknamed Warrior) and his pickaxe (Warlord), exposing a number of new *Triceratops* bones. As he and the others pulverized the hillside, more bones were jarred loose. Eventually the excavation site extended for some seven hundred square feet (sixty-four square meters), and yielded over 130 bones.

It quickly became very complex, so Scott tasked me with making a map—a skill I had learned the previous month from Paul Sereno. I laid out a meter-by-meter grid of string attached to chisels pounded into the rock. Using the grid for reference, I sketched the location of each bone in my field notebook. On the adjacent page I identified each bone, assigned it a number, and made notes on its size and orientation. In this way, we began to make order from the chaos.

The map and bone inventory revealed something peculiar. There were three copies of the same bone: three left nasals, the bone that makes up the front and side of the snout. Each *Triceratops* had only one left nasal, the same way it had only one head or one brain. Then it dawned on us: we had three *Triceratops*es: not only Homer, but Bart and Lisa too. Helmuth had found a *Triceratops* graveyard.

It was the first time that anybody had found more than one *Triceratops* in the same place. Until Helmuth walked into that gully, we thought *Triceratops* was a solitary animal—and we were fairly confident, because *Triceratops* was so common, already known from hundreds of fossils found over a hundred years, each one a single individual, encountered on its own. But one discovery can change everything, and because of what Helmuth found, we now think that *Triceratops* was a pack species.

It's actually not too surprising, because there is ample evidence that close cousins of *Triceratops*—some of the other large, horned ceratopsian species living in other parts of North America during the final 20 million years of the Cretaceous—were social creatures that cohabited in big groups. One of these species, *Centrosaurus*, which lived in modern-day Alberta about 10 million years before *Triceratops* and had a giant horn rising from its nose, has also been found in a bone bed—not a modest bonebed like the Homer site, but one covering an area of nearly three hundred football fields and entombing more than a thousand individuals. Several other ceratopsians have also been found in mass graves, providing a wealth of circumstantial evidence that these big, slow, horned, plant-munching species were communal. It brings to mind an evocative image: these dinosaurs probably moved across Late Cretaceous western North America in vast herds, many thousands strong, rumbling the ground and kicking up clouds of dust as they plowed across the landscape, not much unlike the bison that would conquer the same plains many millions of years later.

After we finished working the Homer site, we continued to prospect the miles of monotonous badlands around Ekalaka, trying to set out early in the morning to beat the worst of the heat. We found a lot of other dinosaur fossils—nothing as important as Homer, but clues from some of the other animals that shared the latest Cretaceous floodplains with *Triceratops* and *T. rex*. We discovered scores of teeth from smaller carnivores, including dromaeosaurid raptors of the *Velociraptor* mold, as well as the chompers of a pony-size animal called *Troodon*, a close relative of the raptors that had developed a taste for a more omnivorous diet. We also came across some foot bones of human-size omnivorous theropods called oviraptorosaurs—weird, toothless dinosaurs with flamboyant crests of bone atop their skulls and sharp beaks adapted to eat a whole variety of food, from nuts and shellfish to plants and small mammals and lizards. Other fossils pointed to two distinct types of herbivores: a fairly boring ornithischian called *Thescelosaurus*, about the size of a horse, and a slightly larger and much more interesting creature called *Pachycephalosaurus*, one of the "dome-headed" dinosaurs with a bowling-ball skull that it used to batter its rivals in fights over mates and territory.

We also spent a couple of days excavating at another locality, which we hoped would turn out to be as productive as the Homer site. It didn't live up to our expectations, but it did produce bones of what is the third most common dinosaur in the Hell Creek formation: another plant-eater called *Edmontosaurus*. At about seven tons in weight and forty feet (twelve meters) from snout to tail, *Edmontosaurus* was a big herbivore like *Triceratops* but of a very different breed. It was a hadrosaur, a member of the duck-billed clan of dinosaurs that evolved

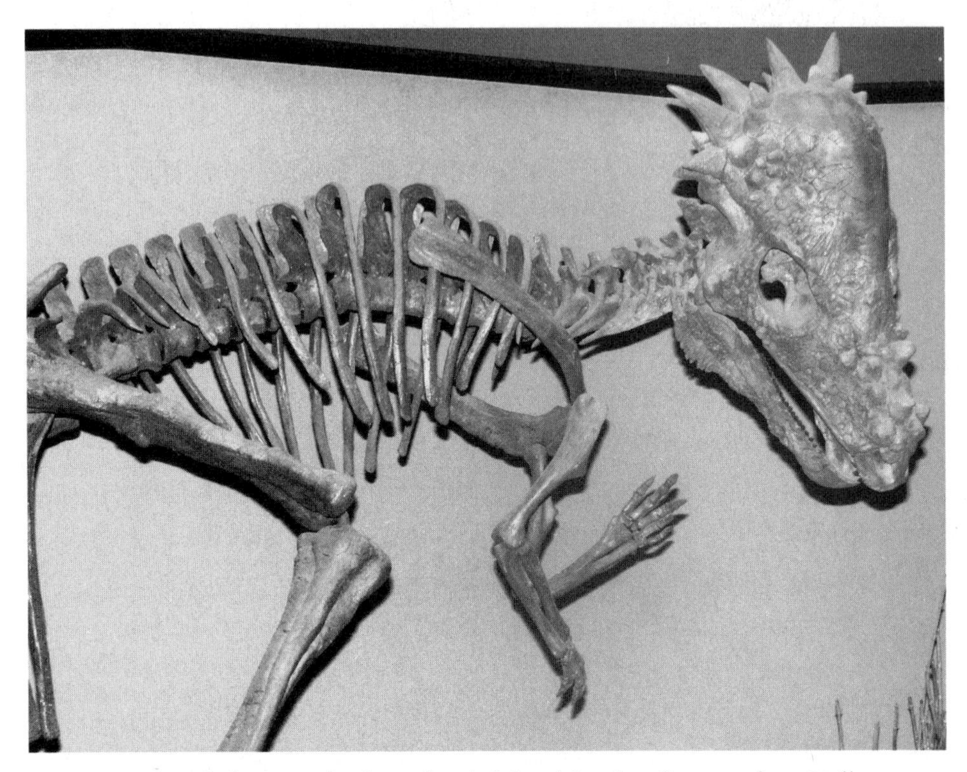

Pachycephalosaurus, the dome-headed, head-butting dinosaur from Hell Creek.

from a separate branch of the ornithischian family tree. They were also very common in the Late Cretaceous—particularly in North America—and many of them lived in herds, walking on either two or four legs depending on how fast they wanted to move, and communicating with bellowing sounds produced by the convoluted spaghetti-twisted nasal chambers within their elaborate head crests. Their nickname comes from the broad, toothless, ducklike bill at the front of their snout, which they used to snare twigs and leaves. Like ceratopsians, their jaws were modified into scissors for slicing—but with even more, and more tightly packed, teeth. Nor were their jaws limited to simple up-and-down movements, but they could pivot from side to side and even hinge outward a little bit, allowing for complex chewing motions. They were some of the most intricate feeding machines ever produced by evolution.

The hadrosaurs, and probably also the ceratopsians, had these sophisticated jaws for a reason. They were fine-tuned by evolution to feed on a new type of plant that had arisen earlier in the Cretaceous: the angiosperms, more commonly known as the flowering plants. Although flowering plants are exceedingly abundant today—the source of much of our food, the décor in many of our gardens—they would have been unknown to the first dinosaurs rising up on Pangea in the Triassic. They were likewise unfamiliar to the giant long-necked sauropods of the Jurassic, which instead inhaled other types of vegetation like ferns, cycads, ginkgos, and evergreen trees. Then, about 125 million years ago, in the Early Cretaceous, small flowers emerged in Asia. With another 60 million years of evolution, these proto-angiosperms had diversified into a range of shrubs and trees, including palms and magnolias, that dotted the Late Cretaceous landscape and that were tasty fodder for the new types of herbivorous dinosaurs that could eat them. There may have even been a little bit of grass—a very specialized type of angiosperm—sprinkled on the ground, but proper grasslands would not develop until much later, many tens of millions of years after the dinosaurs cleared out.

Hadrosaurs and ceratopsians eating flowers. Smaller ornithischians feeding on shrubs, the pachycephalosaurs head-butting each other in tests of dominance. Poodle-size raptors prowling for salamanders, lizards, even some of our early mammal relatives, all of which are known from Hell Creek fossils. A variety of omnivores—*Troodon* and the freakish oviraptorosaurs—picking up whatever scraps the more specialized meat-eaters and plant-eaters forgot about. Other dinosaurs I haven't yet mentioned, like the speed-demon ornithomimosaurs, and the heavily armored *Ankylosaurus*, fighting for their own niches. Pterosaurs and primitive birds soaring overhead; crocodiles lurking offshore in the rivers and the lakes. Not a sauropod to be found, and the King—the great *T. rex* itself—ruling over all of it.

This was the Late Cretaceous of North America, the final flourish of the dinosaurs before disaster struck. Because of the wealth of fossils discovered by everyone from Barnum Brown to the teams from the Burpee Museum, it is the single richest dinosaur ecosystem known to science during the entire Age of Dinosaurs anywhere in the world, our best picture of how a variety of dinosaurs lived together and fit together into one food chain.

It was much the same story in Asia, where big tyrannosaurs like my Pinocchio rex reigned over communities of duckbills, domeheads, raptors, and theropod omnivores—due to the close physical proximity with North America that allowed regular exchange of species between the two continents.

Meanwhile, south of the equator, things were much different.

ALMOST SMACK IN the middle of Brazil is a gently rolling plateau that was once covered by woodland savannah but is now prime farming country. There people grow some of the same crops found in the fields that stretch between my hometown and the Burpee Museum—mostly corn and soybeans—but also more exotic things like sugarcane, eucalyptus, and a whole host of delicious but unfamiliar fruits. This area is called Goiás, and it's a landlocked state of some six million residents, crisscrossed by lonely highways. The national capital, Brasília, is a few hours away, and the Amazon surges a thousand miles to the north. Few foreign tourists ever make it here.

Goiás, however, holds many secrets. You wouldn't know it from the mundane topography, but underneath the farms is a hidden landscape, one that was on the surface between 86 and 66 million years ago. It is a terrain of windblown deserts on the fringes of great river valleys, represented today by a thousand-foot-thick basement of rocks, the foundation for the corn and bean fields. These rocks were molded out of the sand dunes, rivers, and lakes of the Late Cretaceous, in what was then a great basin formed from the residual stresses of South America and Africa cracking apart. This basin was a haven for dinosaurs.

The Cretaceous rocks of Goiás remain mostly buried, but they do poke up here and there, along roads or stream banks. The best place to see them, though, is in quarries, where heavy machinery has torn through the earth to expose the layers of sandstone and mudstone beneath. That's where I found myself one day in early July 2016, the beginning of the austral winter but still hot and muggy, adorned with a hard hat to save my scalp from falling stone and shin guards up to my knees to protect against an even greater danger: snakes. I had been invited to Brazil by Roberto Candeiro, a professor at the Universidade Federal de Goiás, the main university in the state, and an expert on the dinosaurs of South America. I had excavated and studied a lot of Late Cretaceous dinosaurs in North America and Asia, but Roberto advised me to get a southern perspective. He didn't mention snakes as part of the deal.

A few years earlier, Roberto had started a new undergraduate geology program at his university's palm-lined campus on the rapidly growing outskirts of Goiânia, the state capital. The bleached white of the lecture halls—whose corridors were open to the breezy subtropical air—contrasted with the dirt streets and aluminum-roofed shanties just a few miles away. Mopeds growled their way through traffic while old men chopped coconuts with machetes on the roadsides, and monkeys swung from the trees in the distance. The next time I return, many of these remnants of old Brazil will probably be gone.

The excitement of the new course, on the sparkling campus in the biggest city around, attracted a number of keen students, some of whom were joining Roberto and me on the trip to the quarry. There was Andre, a vivacious, potbellied comedian going back to school after trying out many different careers—papaya grower, taxi driver, and years ago a ranch hand in charge of manually deseminating male, and artificially inseminating female, pigs on one of those big farms in the flatlands. Much younger was eighteen-year-old Camila, a short wisp of a woman whose stature belies her boundless energy and ferocity—she relieves stress by kickboxing in her spare time. And then Ramon, a tall, tanned heartbreaker who, with his skinny jeans and hair slicked over to one side, could have leaped right out of one of the Brazilian boy-band music videos that seemed to be playing on every restaurant television.

The quarry we were gathered in was owned by a young guy whose family had been farming in central Brazil for generations. They mined the rock for fertilizer. It is a strange type of stone that looks like concrete, with pebbles of various shapes and sizes embedded within a white matrix. The white stuff is limestone; the pebbles are various rocks that were washed around by the

Roberto Candeiro searching for fossils in Goiás, Brazil.

raging rivers of latest Cretaceous Brazil. Among those pebbles are rare bones—dinosaur fossils. Maybe one out of every ten or twenty thousand of them is bone instead of boulder, but whatever bones you can find are treasure, because they are the remains of some of the last dinosaurs of South America, those species living around the same time as *T. rex*, *Triceratops*, and the Hell Creek gang up north.

Alas, after many hours of searching, we didn't find any bones in the quarry when I visited. We also didn't get bitten by any snakes, so it was a rare day when I came home from the field empty-handed but happy. Later during the trip, we did find some bones in other places, but only fragments. There would be no new species this time—which is often the case when you're exploring a new area, because finding totally new dinosaurs is a hard job, dependent on luck and circumstance. But Roberto has led many such field trips over the past decade, often taking along his motley crew of students, and they've found a lot of bones. Roberto keeps some of them in his lab in Goiânia, where I spent the remainder of my time in Brazil working with Roberto and another of his buddies, an oil-company geologist named Felipe Simbras, who studies dinosaurs as a hobby.

When you look at the fossils shelved in Roberto's lab, it's striking to see no *T. rex*. No tyrannosaurs of any kind, in fact, are known from the latest Cretaceous of Brazil. Spend a day walking through the Hell Creek badlands in Montana, and you'll probably find several *T. rex* teeth—they're that common. But zilch in Brazil, or anywhere else in the southern half of the planet. Instead, Roberto has drawers of other types of carnivorous dinosaur teeth. Some of these belong to a group that we've already met: the carcharodontosaurs, that clan of mighty meat-eaters that evolved from the allosaurs and terrorized much of the planet earlier in the Cretaceous. A few of them, like *Carcharodontosaurus* from Africa, which I studied with Paul Sereno, eventually reached sizes rivaling *T. rex*. Up north, the carcharodontosaurs came and went, ruling for tens of millions of years before ceding their crown to the tyrannosaurs in the middle Cretaceous. Down south they persisted to the very end of the Cretaceous, retaining their heavyweight title because there were no tyrannosaurs around to take it.

Another type of tooth is commonly found in Brazil. They're also sharp, serrated blades, so they must have come from the mouth of a carnivore, but they are usually a little smaller, more delicate. They belong to a different group of theropods called abelisaurids, an offshoot of fairly primitive Jurassic stock that found the southern continents ripe for the taking during the Cretaceous. A decent skeleton of one, called *Pycnonemosaurus*, was found one state over from Goiás, in Mato Grosso. The bones are fragmented but are thought to belong to an animal that was about thirty feet (nine meters) long and weighed a couple of tons.

Even better skeletons of abelisaurids have been found farther south, in Argentina, while others have been discovered in Madagascar, Africa, and India. These more complete fossils—*Carnotaurus*, *Majungasaurus*, and *Skorpiovenator* among them—reveal abelisaurids as fierce animals, a little bit smaller than tyrannosaurs and carcharodontosaurs, but still at or near the top of the food chain. They had short, deep skulls, sometimes with stubby horns jutting out from near the eyes. The bones of the face and snout were encrusted with a rough, scarred texture, which probably supported a sheath made of keratin. They walked on two muscular legs like *T. rex*, but had even more pitiful arms. Although it was thirty feet (nine meters) long and 1.6 tons in weight, *Carnotaurus* had arms barely bigger than a kitchen spatula, which flopped around in a useless way, probably all but invisible if you were watching it go about its everyday business. Clearly the abelisaurids didn't need their arms, relying on their jaws and their teeth for all of the dirty work.

That dirty work, for both abelisaurids and carcharodontosaurids, was catching and chomping the other dinosaurs they lived with, particularly the plant-eaters. Some of them were similar to northern species—for example, some duck-billed dinosaurs have been found in Argentina. But for the most part, it was a different bunch of herbivores down south. There were no pulsating herds of ceratopsians like *Triceratops*, and no dome-headed pachycephalosaurs. There were, however, sauropods. Hordes of them. *T. rex* didn't chase down any of these long-necked titans up in ancient Montana, as sauropods seemed to have disappeared from most of North America some time during the middle part of the Cretaceous (although they still did frequent the southern reaches of the continent). Not so in Brazil or the other austral lands. There sauropods remained the primary large-bodied plant-eaters, right up to the end of the Age of Dinosaurs.

It was one particular type of sauropod that spread across the southlands. The halcyon days of the Jurassic were far gone, and no longer did *Brachiosaurus*, *Brontosaurus*, *Diplodocus*, and their ilk crowd together in the same ecosystems, finely dividing the niches between them with their distinctive teeth, necks, and feeding styles. What was left at the end of the Cretaceous was a more restricted roster of sauropods, a subgroup called the titanosaurs. Some were truly Biblical in proportions—like *Dreadnoughtus* from Argentina or *Austroposeidon*, described by Felipe the oilman and his colleagues from a series of vertebrae—each one the size of a bathtub—found directly south of Goiás in São Paulo State. It's the largest dinosaur ever found in Brazil, and probably stretched about eighty feet (twenty-five meters) from snout to tail. It strains the senses to envision what it must have weighed, but probably somewhere in the ballpark of twenty to thirty tons, maybe much more.

Other late-surviving southern titanosaurs—from Brazil and elsewhere—were considerably smaller. The so-called aeolosaurins were modest creatures, at least as sauropods go, with some of the better-known species, like *Rinconsaurus*, at a mere four tons and thirty-six feet (eleven meters) long. Another subgroup, called saltasaurids, were of the same general size, and they protected themselves from the hungry abelisaurids and carcharodontosaurs with a patchwork of armor plates implanted in their skin.

We also know that there were some smaller theropods but nothing like the panoply of small to midsize carnivores and omnivores in North America. Maybe, you could argue, we just haven't found their small and delicate bones yet, but that's not a very satisfying explanation, because there are many skeletons of similar-size animals found in Brazil, but they are crocodiles, not theropods. Some of them were fairly standard water dwellers that probably wouldn't have competed very much with dinosaurs, but others were bizarre animals adapted for living on the land, so unlike today's crocs. *Baurusuchus* was a long-legged, doglike pursuit predator. *Mariliasuchus* had teeth that looked like the incisors, canines, and molars of mammals, which it probably used like pigs to eat a smorgasbord omnivorous diet. *Armadillosuchus* was a burrower with bands of flexible body armor, and it may have been able to roll up in the style of an armadillo, hence its name. None of these animals lived in North America, as far as we know. It seems that, in Brazil and throughout the Southern Hemisphere, these crocodiles were filling ecological niches held by dinosaurs in other parts of the world.

Carcharodontosaurs and abelisaurids instead of tyrannosaurs, sauropods instead of ceratopsians, swarms of crocs instead of raptors, oviraptorosaurs, and other small theropods. The north and the south were different from each other during those waning years of the Cretaceous, that much is certain. But these big continental areas were downright normal—boring, even—compared to what was going on at the same time in the middle of the Atlantic, where some of the weirdest dinosaurs to ever evolve were hopping around the flooded remnants of Europe.

OF ALL THE people who have ever studied dinosaurs, collected dinosaur bones, or even thought about dinosaurs in any serious way, there's never been anybody quite like Franz Nopcsa von Felső-Szilvás.

Baron Franz Nopcsa von Felső-Szilvás, I should say, because this man was literally an aristocrat who dug up dinosaur bones. He seems like the invention of a mad novelist, a character so outlandish, so ridiculous, that he must be a trick of fiction. But he was very real—a flamboyant dandy and a tragic genius, whose exploits hunting dinosaurs in Transylvania were brief respites from the insanity of the rest of his life. Dracula, in all seriousness, has nothing on the Dinosaur Baron.

Nopcsa was born in 1877 to a noble family in the gentle hills of Transylvania, in what is now Romania but was then on the fringes of the decaying Austro-Hungarian Empire. He spoke several languages at home, and they instilled within him an urge to wander. He also had urges of another kind, and when he was in his twenties, he became the lover of a Transylvanian count, an older man who regaled him with tales of a hidden kingdom of mountains to the south, where tribesmen wore dapper costumes, brandished long swords, and spoke in an indecipherable tongue. The local mountain men called their homeland Shqipëri.

We know it today as Albania, but then it was a backwater on the southern edge of Europe, occupied for centuries by another great empire, the Ottoman.

The baron decided to check it out for himself. He headed south, through the borderlands that separated two empires, and when he arrived in Albania, he was welcomed with a gunshot, which sliced through his hat and narrowly missed his skull. Undeterred, he proceeded to cross much of the country on foot. He picked up the language, grew his hair long, started dressing like the natives, and earned the respect of the insular tribes nestled among the mountain peaks. But the tribesmen might not have been so welcoming if they'd known the truth: Nopcsa was a spy. He was being paid by the Austro-Hungarian government to provide intelligence on their Ottoman neighbors, a mission that became even more critical—and dangerous—as the empires collapsed and the map of Europe was redrawn in the hellfires of World War I.

That's not to say that the baron was merely a mercenary. He was enamored of Albania—obsessed, really. He became one of Europe's leading experts on Albanian culture and came to truly love its people—one in particular. Nopcsa fell for a young man from a sheepherding village in the high mountains. This man—Bajazid Elmaz Doda—nominally became Nopcsa's secretary, but he was so much more, although it wasn't spoken about so openly in those less accepting times. The two lovers would remain together for nearly three decades, enduring the leers of their peers, surviving the disintegration of their respective empires, traveling Europe by motorcycle (Nopcsa on the bike, Doda in a sidecar). Doda was by Nopcsa's side when, in the chaos before the Great War, the baron plotted an insurgency of mountain men against the Turks—even smuggling in firearms to build an arsenal—and then later tried to install himself as king of Albania. Both schemes failed, so Nopcsa turned to other pursuits.

As it turned out, that would be dinosaurs.

In fact, Nopcsa became interested in dinosaurs before he knew anything of Albania, before he met Doda. When he was eighteen, his sister picked up a mangled skull on the family estate. The bones had turned to stone, and it didn't look like any animal the young baron had ever seen scurrying or soaring across his stately grounds. He brought it with him when he started university in Vienna later that year, and upon showing it to one of his geology instructors, he was told to go find more. And so he did, obsessively exploring the fields, hills, and riverbeds of the land he would later inherit, on foot and horseback. Four years later, a blueblood in name but still just a student, he stood up in front of the learned men of the Austrian Academy of Sciences and announced what he had been up to and what he had found: a whole ecosystem of strange dinosaurs.

Nopcsa continued to collect Transylvanian dinosaurs for much of the rest of his life, taking breaks here and there when his services were needed in Albania. He studied them, too, and in doing so was one of the first people who made any attempt to grasp what dinosaurs were like as real animals, not simply bones to be classified. He had a genius when it came to interpreting fossils, and it didn't take him very long to notice that something was odd about the bones he was finding on his estate. He could tell that they belonged to groups that were common in other parts of the world—a new species that he named *Telmatosaurus* was a duckbill, a long-necked critter called *Magyaro-*

saurus was a sauropod, and he also found the bones of armored dinosaurs. However, they were smaller than their mainland relatives, in some cases, astoundingly so; while its cousins were shaking the Earth with their thirty-ton frames in Brazil, *Magyarosaurus* was barely the size of a cow. At first Nopcsa thought the bones belonged to juveniles, but when he put them under a microscope, he realized that they had the characteristic textures of adults. There was only one suitable explanation: these Transylvanian dinosaurs were miniatures.

This raised an obvious question: why were they so tiny? Nopcsa had an idea. Along with his expertise in espionage, linguistics, cultural anthropology, paleontology, motorbiking, and general scheming, the baron was also a very good geologist. He mapped the rocks that held the dinosaur fossils and could tell that they had formed in rivers—thick sequences of sandstones and mudstones that were deposited either in the channels or off to the side when the rivers flooded. Underneath these rocks were other layers that came from the ocean—fine clays and shales bursting with microscopic plankton fossils. Tracing out the aerial extent of the river rocks and scrutinizing the contacts between the river and ocean layers, Nopcsa realized that his estate used to be part of an island, which emerged from the water some time during the latest Cretaceous. The mini-dinosaurs were living on a small bit of turf, probably around thirty thousand square miles (eighty thousand square kilometers) in area, about the size of Hispaniola.

Maybe, Nopcsa conjectured, the dinosaurs were small *because* of their island habitat. It stemmed from an idea that some biologists of the time were beginning to entertain, based on studies of modern species living on islands and the discovery of some strange small mammal fossils in the middle of the Mediterranean. This theory held that islands are akin to laboratories of evolution, where some of the normal rules that govern larger landmasses break down. Islands are remote, so it is always a little bit random as to which species can make their way out to them, being carried by the wind or rafting in on floating logs. There is less space on islands, so fewer resources, so some species may not be able to get so big. And, because islands are severed from the mainland, their plants and animals can evolve in splendid isolation, their DNA cut off from that of their continental cousins, each inbred island-living generation becoming more different, more peculiar over time. This, Nopcsa, thought, is why his island-dwelling dinosaurs were so tiny, so funny looking.

Later research showed that Nopcsa was correct, and his dwarf dinosaurs are now regarded as a prime example of the "island effect" in action. Otherwise, fate wasn't so kind to the baron. Austria-Hungary was on the losing side in the Great War, and Transylvania was handed over to one of the winners, Romania. Nopcsa lost his lands and his castle, and a senseless attempt to reclaim his estate ended with him getting pummeled by a gang of peasants and left for dead by the side of the road. With little money to support his lavish lifestyle, Nopcsa grudgingly accepted the directorship of the Hungarian Geological Institute in Budapest, but bureaucratic life was not for him, so he quit. He sold off his fossils and moved to Vienna with Doda, destitute and overcome with a melancholy that we would probably today recognize as depression. Eventually he had enough. In April 1933, the erstwhile baron slipped some sedative into his lover's tea. When Doda drifted off to sleep, Nopcsa put a bullet into him, then turned the gun on himself.

Nopcsa's tragic demise left one mystery. The baron had

cracked the riddle of the island dinosaurs, and he knew why they were small, but almost every bone he found—whether sauropod, duckbill, or armored ankylosaur—came from a plant-eater. He had little clue as to what predators prowled his miniature menagerie. Were there freakish versions of tyrannosaurs or carcharodontosaurs that ruled the island, perhaps ones that skipped over from the continents? Other types of meat-eaters, also of diminutive stature? Or maybe there were no carnivores at all—the herbivores able to shrink in size because there was nothing out there hunting them.

Solving this problem took a century and another remarkable character, a Transylvanian cut from the same cloth as Nopcsa. Mátyás Vremir is also a polymath, a man of many languages, a traveler who sets out for strange lands with little more than his rucksack. He's never been a spy—as far as I know—but for many years he hopscotched around Africa, working on oil rigs and scouting new drill sites. Now he runs his own company in his native city of Cluj-Napoca, doing environmental surveys and geological consulting on building projects. He's also into many other things: skiing and exploring caves in the Carpathians, canoeing the Danube Delta, and rock climbing, often bringing along his wife and two young sons (in this custom, he departs from Nopcsa). Tall and wiry, with the long hair of a rocker and the piercing eyes of a wolf, he has an intense personal code of honor and does not suffer fools gladly—or really, at all—but if he likes and respects you, he will go to war with you. He's one of my favorite people in the world. If I ever found myself in any real danger, in any godforsaken corner of the planet, he's the one person whom I would want by my side, a man I know I could trust with my life.

He has many talents, but what Mátyás does best is find dinosaurs. Along with my friend Grzegorz from Poland, who found all of those footprints of the first dinosauromorphs, Mátyás has the best nose for fossils of anybody I've ever known. And he seems to do it so effortlessly; when we're together in Romania, me all festooned in my pricey field gear and Mátyás strolling along in his board shorts, cigarette dangling from his lips, it's always he who sees the good fossils. But it isn't really that easy. Mátyás is in fact ruthless: when on the scent of fossils, he'll wade into frigid rivers in the Romanian winter, abseil down hundred-foot cliffs, or contort himself into the tightest and deepest of caves. Once I saw him push his way through rapids on a broken foot because he saw a bone sticking out of the opposite riverbank.

At that very same river, in autumn 2009, Mátyás made the most important discovery of his life. He was out prospecting with his boys when he saw some chalk-white lumps poking out from the rusty red rocks on the bank a few feet above the waterline. Bones. He took out his tools and scratched into the soft mudrock, and more kept coming: the limbs and torso of a poodle-size critter. Excitement quickly turned to fear: the local power station would soon be discharging a surge of water into the river, and the rising currents would probably wash away the bones. So Mátyás worked quickly, but with the precision of a surgeon, and cut the skeleton out of its 69-million-year-old tomb. He brought it back to Cluj-Napoca, made sure it was kept safe in the local museum, and then got down to trying to figure out what it was. He was pretty sure it was a dinosaur, but nothing like it had been found in Transylvania before. Some outside advice would be useful, so Mátyás e-mailed a paleontologist who had excavated and described a great variety of small Late Cretaceous dinosaurs: Mark Norell, the dinosaur curator at the American Museum of Natural History in New York, the guy with Barnum Brown's old job.

Like me, Mark gets a lot of random e-mails from people asking him to identify their fossils, which often are nothing more than misshapen rocks or lumps of concrete. But when he opened the e-mail from Mátyás and downloaded the photos that were attached, Mark was gobsmacked. I know because I was there. I was Mark's PhD student at the time, writing a thesis on theropod dinosaur genealogy and evolution. Mark called me into his office—a stately suite looking out over Central Park—and asked what I thought about the cryptic message he had just received from Romania. We both agreed the bones looked like a theropod's, and when we did a bit of research, we realized that no good meat-eating dinosaur skeletons had ever been found in Transylvania. Mark replied to Mátyás and they struck up a friendship, and a few months later, the three of us found ourselves together in the February chill of Bucharest.

We convened in the wood-paneled office of one of Mátyás's colleagues, a thirty-something professor named Zoltán Csiki-Sava, who, after the fall of Communism put an end to his forced conscription in Ceaușescu's army, went to college and became one of Europe's top dinosaur experts. All of the bones were laid out before us on a table, and it was up to us four to identify them. Seeing the specimen with our own eyes, we had no doubt it was a theropod. Many of its light, delicate bones resembled those of *Velociraptor* and other lithe, fierce raptor species. It was about the same size as *Velociraptor*, too, or maybe a tad smaller. But something didn't quite fit. Mátyás's dinosaur had four big toes

Mátyás Vremir surveying the Red Cliffs in Transylvania, on the lookout for fossils of dwarfed dinosaurs.

on each foot, the two inner ones bearing huge, sickle-shaped claws. The raptors were famous for their retractable sickle claws—which they used to slash and gut their prey—but they had only one on each foot. Besides, they had only three main toes, not four. We were stuck in a quandary, and it seemed that we might have a new dinosaur on our hands.

Over the course of the week, we kept studying the bones, measuring and comparing them to the skeletons of other dinosaurs. Finally it dawned on us. This new Romanian theropod was a raptor, but a peculiar one, with extra toes and claws compared to its mainland relatives. This was quite the revelation: while the plant-eating dinosaurs of the ancient Transylvanian island got small, the predators went weird. It wasn't just the double set of killer claws and the extra toe. The Romanian raptor was stockier than *Velociraptor*, many of the bones of its arms and legs were fused together, and it had even withered its hand into a conjoined mass of stubby fingers and wristbones. It was a new breed of meat-eating dinosaur, and a few months later we gave it a fitting scientific name: *Balaur bondoc*; the first word is an archaic Romanian term for dragon and the second means "stocky."

Balaur bondoc was the top dog of the Late Cretaceous European islands. Less tyrant than assassin, *Balaur* would employ its arsenal of claws to subdue the cow-size sauropods and mini-duckbills and armored dinosaurs marooned in the middle of the rising Atlantic. As best we can tell, it was the largest carnivorous dinosaur on the islands. Who knows what fossils Mátyás will find next, but it seems fairly certain he'll never come across a giant tyrannosaurish carnivore. After a century of searching, after the collection of thousands of fossils—of not only bones

The foot of *Balaur*, the tiny top predator of the latest Cretaceous Transylvanian island. *Photograph by Mick Ellison.*

but also eggs and footprints, and not only dinosaurs but also lizards and mammals—not a single scrap of a big flesh-eater has ever turned up. Not even a tooth. That absence is probably telling us something: the island was too small to support giant bone-crunching monsters, so it was the feisty little guys like

Balaur that topped the food chain—another sign of just how unusual these most stupendous of dinosaur ecosystems were during the closing years of the Cretaceous.

ON ONE OF my trips to Transylvania, we took the afternoon off from fossil hunting and headed into the hills. Mátyás stopped the car outside a castle, near a small village called Săcel. It must have been grand once, but now it was falling to ruin, abandoned long ago. Most of the bright green paint on the outside had faded, exposing the bricks. The windows were all busted, the wooden floors were decaying, and the plaster was sprayed with graffiti. Feral dogs wandered about like zombies. Dust clung to every surface. But somehow, as if defying the laws of gravity and the ravages of time, a gilded chandelier hung proudly from the ceiling in the foyer. We walked underneath it, nervously, as we climbed a set of creaking stairs. Upstairs, more squalor was spread before us: an echoing chasm of a room, with a gaping hole where there used to be a bay window.

It was here—a hundred years ago, when it was a library—where Baron Nopcsa sat and read about dinosaurs, learning the nuances of their bones, theorizing about why the fossils he was finding on the grounds outside were so strange. This castle was Nopcsa's home, the seat of his family's dynasty for centuries. Many generations of Nopcsas lived here, and when the baron himself was at the height of his achievement—when he was spying on the Albanians for his empire and lecturing about dinosaurs to packed audiences all over the continent—it probably seemed that many generations more would follow.

So, too, it was with the dinosaurs. Toward the end of the

Cretaceous—when *T. rex* and *Triceratops* were fighting in North America, carcharodontosaurs were hunting gigantic sauropods throughout the south, and a parade of dwarfs had colonized the European islands—dinosaurs seemed invincible. But like castles, like empires, and like genius noblemen with a flair for the dramatic, the great dynasties of evolution can also fall—sometimes when least expected.

8

DINOSAURS TAKE FLIGHT

Archaeopteryx

THERE IS A DINOSAUR OUTSIDE my window. I'm watching it as I write this.

Not a photo on a billboard, or a copy of a skeleton from a museum, or one of those obnoxious animatronic things you see in amusement parks.

A real, honest-to-goodness, living, breathing, moving dinosaur. A descendant of those plucky dinosauromorphs that emerged on Pangea 250 million years ago, part of the same family tree as *Brontosaurus* and *Triceratops*, and a cousin of *T. rex* and *Velociraptor.*

It's about the size of a house cat, but with long arms tucked against its chest, and a much shorter pair of twiggy legs. Most of its body is the crisp white of a bridal gown, but the edges of its arms are gray, and the tips of its hands are jet black. As it stands stiff-legged on my neighbor's rooftop, its head arching proudly upward, it cuts a regal profile against the darkening clouds of eastern Scotland.

When the sun breaks through for a moment, I catch a glint reflecting from its beady eyes, which start to dart back and forth. No doubt this is a creature of keen senses and high intelligence, and it's onto something. Maybe it can tell that I'm watching.

Then, without warning, it yawns open its mouth and emits a high-pitched screech—an alarm to its compatriots, perhaps, or a mating call. Or maybe a threat directed my way. Whatever it is, I can hear it clearly through the double glazing, thankful now that there is a pane of glass between us.

The fluffy-coated critter becomes silent again and swivels its neck so that it's now staring directly at me. It definitely knows I'm here. Expecting another shriek, I'm surprised when it closes its mouth, its jaws coming together to form a sharp, yellow beak, which hooks downward at the front. It doesn't have any teeth, but this beak looks like a nasty weapon that could do a lot of damage. Mindful again that I'm indoors and safe from any harm, I give the glass a playful little tap.

And then the creature makes its move. With a grace that I can only struggle to describe, it pushes its webbed feet off the slate tiles, extends its feathered arms outward, and leaps into the breeze. I lose sight of it as it disappears over the trees, probably on its way to the North Sea.

THE DINOSAUR I'M watching is a seagull. There are thousands of them living around Edinburgh. I see them every day, sometimes diving for fish in the sea a couple of miles north of my house, but more often I watch in disgust as they pick at discarded burger wrappers and other waste on the streets of the Old Town. Occasionally I catch one of them dive-bombing an unsuspecting tourist, spearing a french fry or two with its beak before launching back into the sky. When I observe this type of behavior—the cunning, the agility, the nastiness—it's easy to see the inner *Velociraptor* in an otherwise forgettable seagull.

Seagulls, and all other birds, evolved from dinosaurs. That makes them dinosaurs. Put another way, birds can trace their heritage back to the common ancestor of dinosaurs, and therefore are every bit as dinosaurian as *T. rex*, *Brontosaurus*, or *Triceratops*, the same way my cousins and I are Brusattes because we trace our lineage back to the same grandfather. Birds are simply a subgroup of dinosaurs, just like the tyrannosaurs or the sauropods—one of the many branches on the dinosaur family tree.

It's a notion that's so important, it bears repeating. *Birds are dinosaurs*. Yes, it can be hard to get your head around. I often get people who try to argue with me: sure, birds might have evolved from dinosaurs, they say, but they are so different from *T. rex*, *Brontosaurus*, and the other familiar dinosaurs that we shouldn't classify them in the same group. They're small, they have feathers, they can fly—we shouldn't call them dinosaurs. On the face of it, that may seem like a reasonable argument. But I always have a quick retort up my sleeve. Bats look and behave a whole lot differently than mice or foxes or elephants, but nobody would argue that they're not mammals. No, bats are just a weird type of mammal that evolved wings and developed the ability to fly. Birds are just a weird group of dinosaurs that did the same thing.

And just so there is no confusion, I'm talking about birds—real, true birds. This has nothing to do with another favorite cast member of the Age of Dinosaurs, the pterosaurs. Often referred to as pterodactyls, these were reptiles that glided and soared through the air on long, skinny wings anchored by a stretched fourth finger (the ring finger). Most were about the size of average birds today, but some had wingspans wider than small airplanes. They originated around the same time as dinosaurs in the Pangean days of the Triassic, and died out with most dinosaurs at the end of the Cretaceous, but they were not dinosaurs, and they were not birds. Instead, they were close cousins of dinosaurs. Pterosaurs were the first group of vertebrates (animals with backbones) to evolve wings and fly. Dinosaurs—in the guise of birds—were the second.

This means that dinosaurs are still among us today. We're so used to saying that dinosaurs are extinct, but in reality, over ten thousand species of dinosaurs remain, as integral parts of modern ecosystems, sometimes as our food and our pets, and in the case of seagulls, sometimes as pests. Indeed, the vast majority of dinosaurs died 66 million years ago, when that latest Cretaceous world of *T. rex* versus *Triceratops*, of the giant Brazilian sauropods and Transylvanian island dwarfs, was plunged into chaos. The reign of the dinosaurs ended and a revolution followed, forcing them to cede their kingdom to other species. But a few stragglers made it through, a few dinosaurs that had what it took to endure. The descendants of these remarkable survivors live on today as birds, the enduring legacy of over 150 million years of dinosaur domination, of a dead empire.

THE REALIZATION THAT birds are dinosaurs is probably the single most important fact ever discovered by dinosaur paleontologists. Although we've learned much about dinosaurs over the past few decades, this is not a radical new idea pushed by my generation of scientists. Quite the opposite: it's a theory that goes back a long way, to the era of Charles Darwin.

The year was 1859. After two decades of sitting around and stewing over the observations he made as a young man sailing the world on the HMS *Beagle*, Darwin was finally ready to go public with his startling discovery: species are not fixed entities; they evolve over time. He even had a mechanism to explain evolution, a process he called natural selection. That November, he laid it all out in the *Origin of Species*.

This is how it works. All populations of organisms are variable in their features. For instance, if you look at a bunch of rabbits in nature, they will have slightly different fur colors, even if they all belong to the same species. Sometimes one of those variations confers a survival advantage—say, darker fur that helps a rabbit camouflage itself better—and because of that, the individuals with that feature have a better chance of living longer and reproducing more. If that variation is heritable—if it can be passed on to offspring—then over time it will cascade throughout the population so that the entire rabbit species is now dark-haired. Dark hair has been naturally selected, and the rabbits have evolved.

This process can even produce new species: if a population is somehow divided and each subset goes its own way, evolving its own naturally selected features until the two subsets are so different that they're unable to reproduce with each other, they have developed into separate species. This process has brought into being all of the world's species over the course of billions of years. It means that all living things—modern and extinct—are related, cousins on one grand family tree.

Elegant in its simplicity, so far-reaching in its implications, today we regard Darwin's theory of evolution by natural selection as one of the fundamental rules underpinning the world as we know it. It's what produced the dinosaurs, what molded them into such a fantastic variety of species that were able to rule the planet for so long, adapting to drifting continents, shifting sea levels, changes in temperature, and the threats from competitors hoping to snatch their crown. Evolution by natural selection is also what produced us, and don't be mistaken, it continues to operate right now, constantly, all around us. It's why we're so worried about superbugs that evolve resistance to antibiotics, why we're always in need of new medicines to stay a step ahead of the bacteria and viruses that will do us harm.

Some folks still dispute the reality of evolution today—and

I won't say any more about that—but whatever disagreements we have now pale in comparison to what was happening in the 1860s. Darwin's book—written in beautiful, accessible prose for public consumption—sparked a fury. Some of society's most cherished notions about religion, spirituality, and humankind's place in the universe suddenly seemed up for debate. Evidence and accusations flew back and forth, and both sides were on the lookout for a trump card. For many of Darwin's supporters, the ultimate proof of his new theory would be "missing links," transitional fossils that capture, like a freeze frame, the evolution of one type of animal into another. These would not only demonstrate evolution in action, but could visually convey it to the public in a way that no book or lecture ever could.

Darwin didn't have to wait long. In 1861, quarry workers in Bavaria found something peculiar. They were mining a type of fine limestone that breaks into thin sheets, which was used at the time for lithographic printing. One of the miners—now nameless to history—split open a slab and found a 150-million-year-old skeleton of a Frankenstein creature inside. It had sharp claws and a long tail like a reptile but feathers and wings like a bird. Other fossils of the same animal were soon found in other limestone quarries that sprinkled the Bavarian countryside, including a spectacular one that preserved nearly the entire skeleton. This one had a wishbone, like a bird, but its jaws were lined with sharp teeth, like a reptile. Whatever this creature was, it seemed to be half reptile, half bird.

This Jurassic hybrid was named *Archaeopteryx*, and it became a sensation. Darwin included it in later editions of the *Origin of Species* as evidence that birds had a deep history which could be explained only by evolution. The strange fossil also caught the eye of one of Darwin's best friends and most vociferous supporters. Thomas Henry Huxley is perhaps best remembered as the man who came up with the term *agnosticism* to describe his uncertain religious views, but in the 1860s he was popularly known as Darwin's Bulldog. It was a nickname he gave himself, because he was unrelenting in his defense of Darwin's theory, taking on anyone—in person or in print—who maligned it. Huxley agreed that *Archaeopteryx* was a transitional fossil, linking reptiles and birds, but he went one step further. He noticed that it bore an uncanny resemblance to another fossil discovered in the same lithographic limestone beds in Bavaria, a small flesh-eating dinosaur called *Compsognathus*. So he proposed his own radical new idea: birds descended from dinosaurs.

Debate continued for the next century. Some scientists followed Huxley; others didn't accept the link between dinosaurs and birds. Even as a deluge of new dinosaur fossils emerged from the American West—the Jurassic Morrison dinosaurs like *Allosaurus* and its many sauropod compatriots, the Cretaceous Hell Creek congregation of *T. rex* and *Triceratops*—there didn't seem to be enough evidence to settle the question. Then, in the 1920s, a book by a Danish artist made the simplistic argument that birds couldn't have come from dinosaurs because dinosaurs apparently didn't have collarbones (which birds fuse into wishbones), and although it may sound a little absurd, that viewpoint held sway until the 1960s (and today we realize that dinosaurs did indeed have collarbones, so the point is moot). As Beatlemania swept the globe, protesters marched for civil rights in the American South, and war raged in Vietnam, the consensus was that dinosaurs had nothing to do with birds. They were just very distant cousins that looked kind of similar.

The feather-covered skeleton of *Archaeopteryx*, the oldest bird in the fossil record.

That all changed in 1969, that tumultuous year of Woodstock. Revolution was afoot, as societal norms and traditions were being challenged throughout the West. That spirit of rebellion also percolated into science, and paleontologists started to see dinosaurs differently. Not as the dim-witted, dull-colored, slow-moving wastes of space that defined a pointless era of prehistory, but as more active, dynamic, energetic animals that ruled their world through talent and ingenuity, creatures that were very similar in many ways to living animals—particularly birds. A new generation—led by an unassuming Yale professor named John Ostrom and his rambunctious student Robert Bakker—

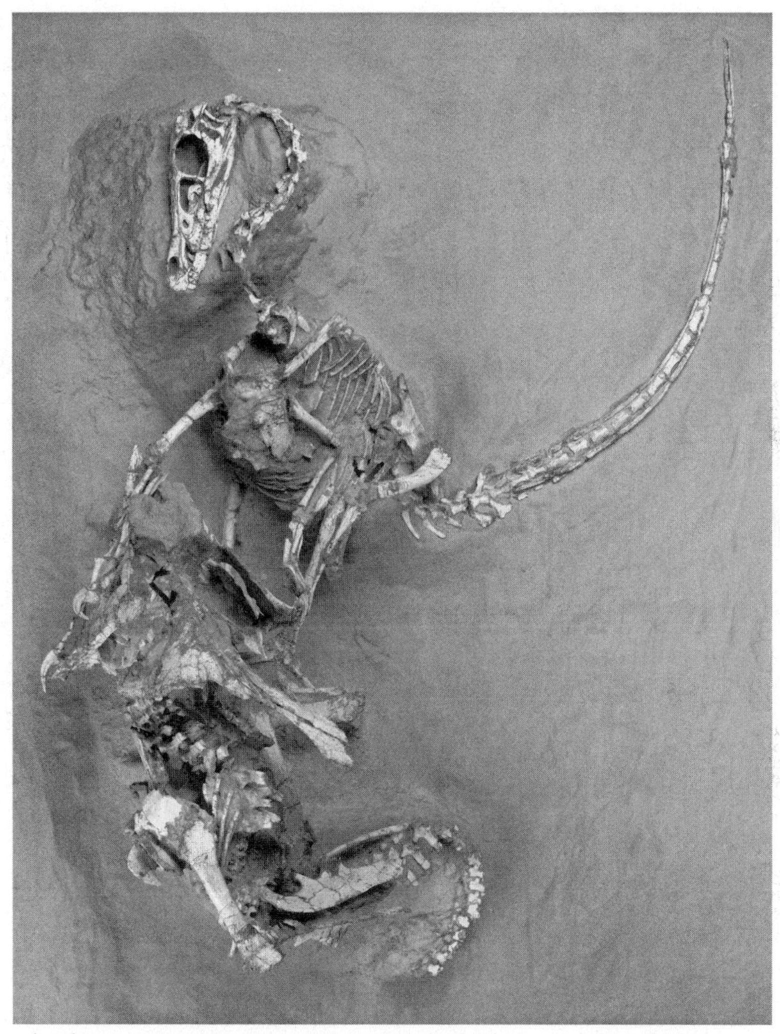

The dromaeosaur (raptor) *Velociraptor* locked in combat with the primitive horned dinosaur *Protoceratops*, from the Gobi Desert of Mongolia. *Photograph by Mick Ellison, with assistance from Denis Finnin.*

completely reimagined dinosaurs, even making the argument that dinosaurs lived together in herds, had keen senses, cared for their young, and may have been warm-blooded like us.

The catalyst for this so-called Dinosaur Renaissance was a

series of fossils unearthed a few years before, in the mid-1960s, by Ostrom and his team. They were out in far southern Montana, close to the border with Wyoming, prospecting in colorful rocks formed on a floodplain during the Early Cretaceous, some time between 125 and 100 million years ago. They found over a thousand bones of a dinosaur—a dinosaur that was astonishingly birdlike. It had long arms that looked almost like wings and the lithe build indicative of a fast-running dynamo of an animal. After a few years of studying the bones, Ostrom announced them in 1969 as a new species: *Deinonychus*, a raptor. It was a close cousin of *Velociraptor*, which was discovered in the 1920s in Mongolia and described by Henry Fairfield Osborn (the New York aristocrat who named *T. rex*), but in these pre–*Jurassic Park* times, it had yet to become a household name.

Ostrom realized the enormous implications of his find. He used *Deinonychus* to resurrect Huxley's idea that birds evolved from dinosaurs, which he argued in a series of landmark scientific papers in the 1970s, a lawyer making his case by meticulous presentation of incontrovertible evidence. Meanwhile, his flamboyant former student Bakker went a different route. The cowboy-hatted, hippie-haired child of the Sixties became an evangelist. He preached the dinosaur-bird connection—and the new image of dinosaurs as warm-blooded, big-brained evolutionary success stories—to the public with a *Scientific American* cover story in 1975 and a wildly successful book in the 1980s, *The Dinosaur Heresies*. Their contrasting styles caused considerable friction between them, but together Ostrom and Bakker revolutionized how everyone viewed dinosaurs. By the end of the 1980s, most serious students of paleontology had come around to their way of thinking.

The recognition that birds came from dinosaurs raised a provocative question. Maybe, Ostrom and Bakker surmised, some of the most familiar features of modern birds first evolved in dinosaurs. Perhaps raptors like *Deinonychus*—so birdy in their bones and body proportions—even had that one thing that is most quintessentially "bird": feathers. After all, because birds evolved from dinosaurs, and because the half-dinosaur half-bird *Archaeopteryx* was found covered in fossilized feathers, feathers must have developed somewhere along their evolutionary lineage—maybe in a dinosaur long before birds came onto the scene. Moreover, if some dinosaurs did have feathers, that would be the final jab in the gut to the few old-blood holdovers who didn't accept the connection between dinosaurs and birds.

The problem, though, is that Ostrom and Bakker couldn't be sure if dinosaurs like *Deinonychus* had feathers. All they had were bones. Soft bits like skin, muscles, tendons, internal organs, and yes, feathers, rarely survive the ravages of death, decay, and burial to become fossilized. *Archaeopteryx*—which Ostrom and Bakker considered the oldest bird in the fossil record—was a lucky exception, having been buried quickly in a quiet lagoon and rapidly turned to stone. Maybe they would never be able to tell one way or another. So they waited, hopeful that somebody, somewhere, somehow would find feathers on a dinosaur.

Then in 1996, as his career was drawing to a close, Ostrom was at the annual meeting of the Society of Vertebrate Paleontology in New York, where fossil hunters from around the world congregate to present new discoveries and discuss their research. While milling about the American Museum, Ostrom was approached by Phil Currie, a Canadian who was part of that first post-1960s generation raised on the idea that birds are dinosaurs. The theory so fascinated Currie that he spent much of the 1980s and 1990s hunting for small birdlike raptors in western Canada, Mongolia, and China. He had, in fact, just returned from one of his trips to China. While he was there, he caught wind of an extraordinary fossil. He took a photograph of it out of his pocket and showed it to Ostrom.

There it was, a small dinosaur surrounded by a halo of feathery fluff, immaculately preserved as if it had died yesterday. Ostrom began to cry. His knees got weak, and he almost fell to the floor. Somebody had found his feathered dinosaur.

The fossil Currie showed Ostrom—later named *Sinosauropteryx*—was only the start. Scientists sprinted to the Liaoning region of northeastern China, where it was found, with the mad ambition of prospectors on a gold rush. But the true authorities were the local farmers. They knew the land intimately and understood that even a single prime specimen, when sold to a museum, could bring them more money than a lifetime of toiling in the fields. Within a few years, farmers from all over the countryside had reported several other feathered dinosaur species, which were given names like *Caudipteryx*, *Protarchaeopteryx*, *Beipiaosaurus*, and *Microraptor*. Today, some two decades later, more than twenty such species are known, and these are represented by thousands of individual fossils. These dinosaurs had the great misfortune to live in a dense forest surrounding a wonderland of ancient lakes, a landscape that was periodically obliterated by volcanoes. Some of these eruptions spewed out tsunamis of ash, which combined with water to flood the landscape in a viscous ooze that buried everything in sight. The dinosaurs were captured going about their everyday business, preserved Pompeii-style. That's why the details of the feathers are so pristine.

Ostrom was a guy who waited hours for a bus, only to have five come along at once. He now had a whole ecosystem of feathered dinosaurs, which proved him right: birds really did arise from dinosaurs, an extension of the same family as *T. rex* and *Velociraptor.* The feathered dinosaurs of Liaoning are now among the most celebrated fossils in the world, and rightly so. When it comes to new dinosaur discoveries, nothing in my lifetime approaches their significance.

ONE OF THE greatest privileges I've had in my career is studying many of the feathered dinosaurs of Liaoning, in museums all over China. I've even had the chance to name and describe a new one, the raptor *Zhenyuanlong* that we met back in the first few pages of this book, that mule-size creature with wings. These Liaoning dinosaurs are gorgeous fossils—as suited for an art gallery as a natural history museum—but they're so much more than that.

They are *the* fossils that help us untangle one of the biggest riddles of biology: how evolution produces radically new groups of organisms, with restyled bodies capable of remarkable new behaviors. The formation of small, fast-growing, warm-blooded, flying birds from ancestors that looked like *T. rex* and *Allosaurus* is a prime example of this sort of jump—what biologists call a major evolutionary transition.

You need fossils to study major transitions, because they're not the sort of thing we can re-create in the lab or witness in nature. The Liaoning dinosaurs are an almost perfect case study. There are a lot of them, and they exhibit great diversity of body size, shape, and feather structure. They run the gamut from dog-size plant-eating ceratopsians with simple porcupine-style quills, to thirty-foot-long primitive cousins of *T. rex* coated in hairlike fuzz (like *Yutyrannus*, which we also met a few chapters ago), to raptors like *Zhenyuanlong* with full-on wings, and even to crow-size weirdos with wings on both the arms and the legs, something not seen in any modern birds. Each is a snapshot, and when stitched together and placed on a family tree, they provide something of a running film of an evolutionary transition in action.

Most fundamentally, the Liaoning fossils confirm where birds perch on the dinosaur family tree. Birds are a type of theropod; they are rooted in that group of ferocious meat-eaters that most famously includes *T. rex* and *Velociraptor* and also many of the other predators that we've come across: the herd-living *Coelophysis* from Ghost Ranch, the Butcher *Allosaurus* from the Morrison Formation, the carcharodontosaurs and abelisaurids that terrorized the southern continents. This is exactly what Huxley, and later Ostrom, proposed. The Liaoning fossils sealed the deal by verifying how many features are shared uniquely by birds and other theropods: not just feathers, but also wishbones, three-fingered hands that can fold against the body, and hundreds of other aspects of the skeleton. There are no other groups of animals—living or extinct—that share these things with birds or theropods: this must mean that birds came from theropods. Any other conclusion requires a whole lot of special pleading.

Among theropods, birds nest within an advanced group called the paravians. These carnivores break some of the stereotypes that many people still hold about dinosaurs, particularly theropods. They weren't lumbering monsters like *T. rex*, but

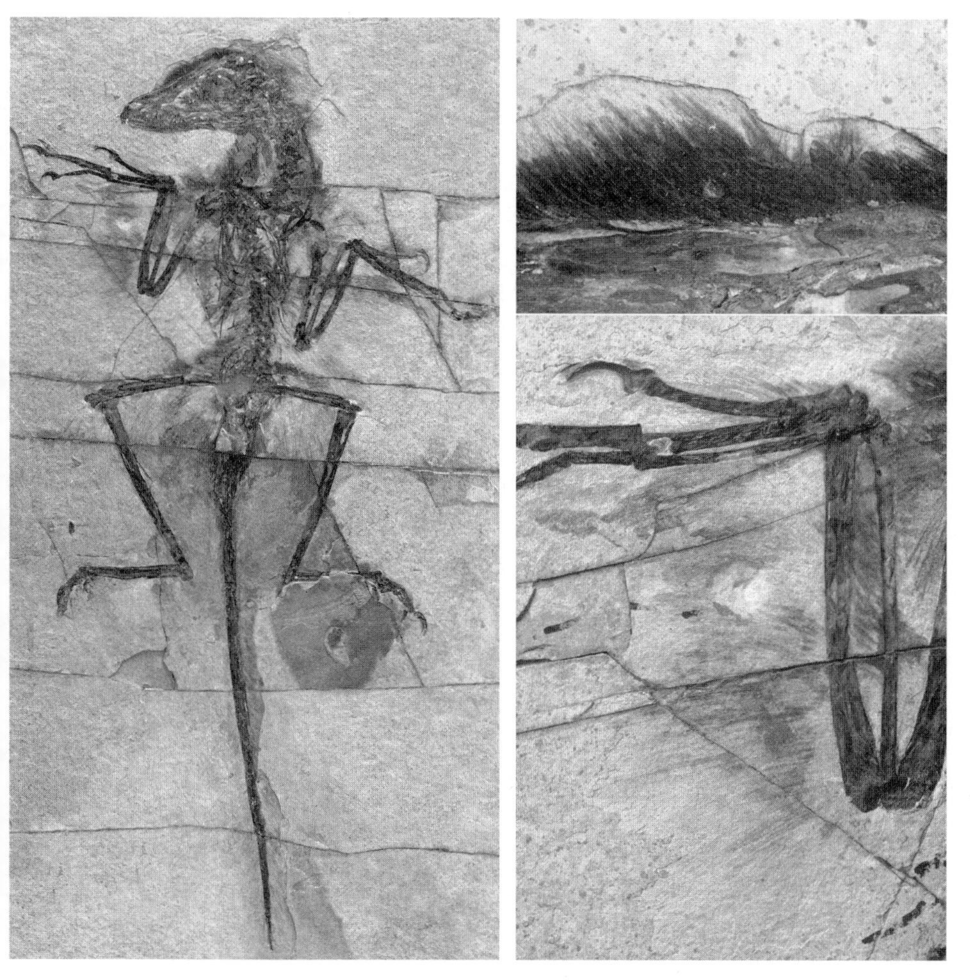

LEFT The feathered dromaeosaur (raptor) *Sinornithosaurus* from Liaoning, China. *Photograph by Mick Ellison.* RIGHT Close-ups of the simple filament-like feathers along the head (top) and longer, quill-type feathers along the forearm (bottom) of *Sinornithosaurus*. *Photograph by Mick Ellison.*

smaller, nimbler, smarter species, most of which were human-size or tinier. In effect, they were a subgroup of theropods that went on their own path, trading the brawn and girth of their ancestors for bigger brains, keener senses, and more compact, lightweight skeletons that permitted a more active lifestyle. Other paravians include Ostrom's *Deinonychus*, *Velociraptor*, and my oh-so-birdlike *Zhenyuanlong*, along with all of the other dromaeosaurid raptor and troodontid species. These dinosaurs are the closest relatives of birds. They all had feathers, many of them had wings, and more than a few surely looked and acted like modern birds.

Somewhere within this flock of paravian species lies the line between non-bird and bird. As with the division between non-dinosaur and dinosaur, way back in the Triassic, this distinction is blurry. And it's getting less distinct with each new fossil from Liaoning. Truthfully, it's just a matter of semantics: today's paleontologists define a bird as anything that falls into the group that includes Huxley's *Archaeopteryx*, modern birds, and all descendants of their Jurassic common ancestor. It's more historical convention than a reflection of any biological distinction. By this definition, *Deinonychus* and *Zhenyuanlong* fall ever so slightly on the non-bird side of the border.

Let's forget about that for a second. Definitions can distract from the story line.

Today's birds stand out among all modern animals. Feathers, wings, toothless beaks, wishbones, big heads that bob along on an S-shaped neck, hollow bones, toothpick legs . . . the list goes on. These signature features define what we call the bird body plan: the blueprint that makes a bird a bird. This body plan is behind the many superskills that birds are so renowned for: their ability to fly, their hypercharged growth rates, their warm-blooded physiology, and their high intelligence and sharp senses. We want to know where this body plan came from.

The feathered dinosaurs of Liaoning give us the answer.

And it's remarkable: many of the supposedly signature features of today's birds—the components of their blueprint—first evolved in their dinosaur ancestors. Far from being unique to birds, these features developed much earlier, in ground-living theropods, for reasons wholly unrelated to flight. Feathers are the best example—and we'll return to them in a moment—but they are merely emblematic of a much bigger pattern. To see it, we have to start at the base of the family tree and move up.

Let's begin with a central feature of the bird blueprint. Long, straight legs and feet with three skinny main toes—hallmarks of the modern bird silhouette—first appeared more than 230 million years ago in the most primitive dinosaurs, as their bodies were reshaped into upright-walking, fast-running engines that could outpace and outhunt their rivals. In fact, these hind-limb features are some of the defining characteristics of all dinosaurs, the very things that helped them rule the world for so long.

Then a little bit later, some of these upright-walking dinosaurs—the earliest members of the theropod dynasty—fused their left and right collarbones into a new structure, the wishbone. It was a seemingly minor change, which stabilized the shoulder girdle and probably allowed these stealthy, dog-size predators to better absorb the shock forces of grabbing prey. Much later, birds would co-opt the wishbone to serve as a spring that stores energy when they flap their wings. These proto-theropods, however, never could have known this would eventually happen, just as the inventor of the propeller had no idea the Wright Brothers would later put it on an airplane.

Many tens of millions of years down the line, a subset of these upright-walking, wishbone-chested theropods called maniraptorans developed a gracefully curved neck, for reasons unknown. I speculate it may have had something to do with scouting for prey. Meanwhile, some of these species were getting smaller in size, probably because their shrinking physiques gave them entry to new ecological niches—trees, brush, perhaps even underground caves or burrows that were inaccessible to giants like *Brontosaurus* and *Stegosaurus*. Later, a subset of these small, upright, wishboned, bobbing-necked theropods started to fold their arms against the body, probably to protect the delicate quill-pen feathers that were evolving around the same time. These were the paravians—a subgroup of the maniraptorans, and the immediate ancestors of birds.

These are just a few examples; there are many more. The point is, when I look at that seagull outside my window, many of the features that allow me to immediately recognize it as a bird are not actually trademarks of birds. They're attributes of dinosaurs.

This pattern isn't confined to anatomy, either. Many of the most notable behaviors and biological characteristics of living birds also have deep dinosaurian heritage. Some of the best evidence comes not from Liaoning but from another trove of spectacular fossils, found in the Gobi Desert of Mongolia. For the last quarter century, a joint team from the American Museum of Natural History and the Mongolian Academy of Sciences has been mounting annual summer expeditions to this desolate expanse of central Asia. The fossils they have collected—which date from the Late Cretaceous, between about 84 and 66 million years ago—provide unprecedented insight into the lifestyles of dinosaurs and early birds.

Leading the Gobi project is one of America's most prominent paleontologists, Mark Norell, the head of the American Museum's

Mark Norell using one of his signature tricks for collecting fossils in damp conditions: drenching the plaster jacket covering the fossil in gasoline and lighting it on fire. *Aino Tuomola.*

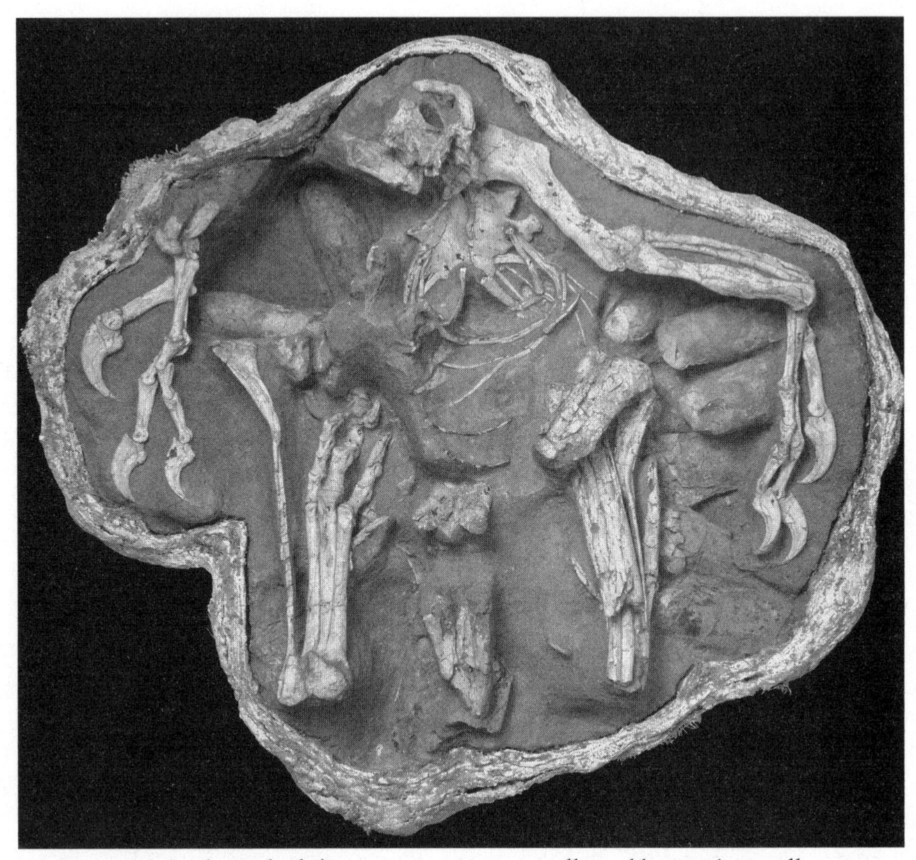

An oviraptor buried while protecting its nest, collected by Mark Norell in Mongolia.

dinosaur collection and my former PhD supervisor. Mark grew up in Southern California, a long-haired surfer dude who worshipped Jimmy Page but at the same time had a nerdy obsession with collecting fossils. He did his graduate work at Yale, where Ostrom was one of his mentors, and was barely into his thirties when he was hired to Barnum Brown's old curatorial post, widely regarded as the top dinosaur research job in the world.

The total opposite of a stuffy academic caricature, Mark travels the planet hunting for the two things he knows best: dinosaurs, obviously, but also his other infatuation: Asian art. The stories he's accumulated along the way—in auction houses, Chinese dance clubs, Mongolian yurts, fancy European hotels, and seedy bars—often seem too outrageous to be true but make him one of the best raconteurs I've known. A few years ago, the *Wall Street Journal* published a hagiography of Mark, referring to him as "the coolest dude alive." Mark does dress like a hipster version of Andy Warhol (another of his heroes), hold court in a majestic office overlooking Central Park, boast a collection of ancient Buddhist art that puts many museums to shame, and bring portable fridges into the desert so he can make sushi while doing fieldwork. Is it enough to qualify as the single coolest individual in the world? I'll let others judge.

I do know that Mark is one of the world's best advisors. He's whip smart and thinks big, always urging his students to ask fundamental questions about how evolution works—for instance, how did a dinosaur turn into a bird? Never a micromanager or a credit stealer, he tries to attract motivated students, supplies them with kick-ass fossils, and then steps aside. That, and he never lets his students pay for beer.

I and many of Mark's students have built our careers studying dinosaurs he wrested from the Gobi. Among them are skeletons entombed by flash storms that captured parent dinosaurs brooding their nests of eggs just like the birds we know today. They show that birds inherited their superb parenting skills from their dinosaurian antecedents, and that these behaviors go back at least to some of the small, winged, arch-necked maniraptoran species. Mark's crews have also discovered a wealth of dinosaur skulls, including the well-preserved crania of *Velociraptor* and other maniraptorans. CAT scanning of these specimens—

spearheaded by Mark's former student Amy Balanoff, who we met a couple of chapters ago—has revealed that these dinosaurs had huge brains, with an expanded forebrain at the front. It is the large forebrain that makes modern birds so intelligent and acts as their in-flight computer, allowing them to control the complicated business of flying and navigating the complex 3-D world of the air. We don't precisely know why these maniraptorans evolved such intelligence, but the Gobi fossils tell us that the ancestors of birds got smart before they took to the skies.

The list continues. Numerous theropods, found in the Gobi and elsewhere, had bones hollowed out by air sacs, which, as we learned earlier, are telltale signs that they had the ultra-efficient "flow through" lung that takes in oxygen during both inhalation and exhalation, a precious feature of birds that delivers the juice needed to maintain their high-energy way of life. The microscopic structure of dinosaur bones indicates that many species—including all known theropods—had growth rates and physiologies intermediate between slow-maturing, cold-blooded reptiles and the fast-growing, warm-blooded birds of today. Thus, we now know that a flow-through lung and relatively fast growth emerged more than 100 million years before birds took wing, as those first fast-running, long-legged dinosaurs were carving out a new livelihood as energetic live wires so different from the sluggish amphibians, lizards, and crocodiles they were battling. We even know that both the typical sleeping posture of birds and the way in which they mine calcium from their bones for the shells of their eggs first arose in dinosaurs long before birds.

What we think of as the bird body plan, therefore, wasn't so much a fixed blueprint as a Lego set that was put together brick by brick over evolutionary time. The same was true of the classic behavioral, physiological, and biological repertoire of today's birds. And the same was true of feathers.

WHENEVER I VISIT China, I always make time to see Xu Xing. He's a polite, mild-tempered man who grew up poor in Xinjiang, a politically disputed sweep of western China that was once crossed by the Silk Road. Unlike most children in the West, Xu had no interest in dinosaurs when he was young. He didn't even know they existed. When he won a prestigious scholarship to go to college in Beijing, the government told him that he would study paleontology, a subject he had never heard of before. Xu complied and actually enjoyed it, and then he went on to train further under Mark Norell in New York. Today, he's the world's greatest dinosaur hunter. He's named more than fifty new species, more than anybody else alive.

Compared to Mark's presidential suite in the turret of the American Museum of Natural History, Xu's office at the Institute of Vertebrate Paleontology and Paleoanthropology in Beijing is spartan. But it contains some of the most amazing fossils you'll ever see. In addition to the dinosaurs that Xu finds himself, he is routinely sent bones that have been collected by farmers, construction workers, and various other people from all over China. Many of those are new feather-covered dinosaurs from Liaoning. Whenever I visit and stride up to Xu's door, I feel the adrenaline of a kid running into a toy store.

The fossils I've seen in Xu's office tell the story of how feathers evolved. More than any other part of a bird's body or biology, feathers are central to figuring out where birds—and many of their unique abilities, like flight—came from. Feathers are nature's ultimate Swiss Army knife, multipurpose tools that can be used for display, insulation, protection for eggs and babies, and of course, flight. Indeed, they have so many uses that it has been difficult to figure out which purpose they first evolved to serve and how they were modified into airfoils, but the Liaoning fossils are starting to provide an answer.

Feathers didn't suddenly spring forth when the first birds entered the scene; they evolved in their distant dinosaurian ancestors. The common ancestor of all dinosaurs may have even been a feathered species. We don't know for sure, because we can't study that ancestor directly, but it's an inference based on an observation: so many small dinosaurs from Liaoning that are well preserved—the bounty of meat-eating theropods like *Sinosauropteryx* but also pint-size plant-eaters like *Psittacosaurus*—are found coated in some type of integument. Either these various dinosaurs evolved their feathers separately, which is unlikely, or they inherited them from a deep ancestor. These earliest feathers, however, looked very different from the quill pens of modern birds. The material that glazed the body of *Sinosauropteryx* and most other Liaoning dinosaurs was more like fluff, made up of thousands of hairlike filaments that paleontologists call proto-feathers. No way could these dinosaurs fly—their feathers were too simple, and they didn't even have wings. So the first feathers must have evolved for something else, probably to keep these small, chinchilla-like dinosaurs warm or maybe as a way to camouflage their bodies.

For most dinosaurs—the vast majority of those that I've studied in Xu's office and in other Chinese museums—a coat of fluffy or bristly feathers was enough. However, in one subgroup—those wishboned, swan-necked maniraptorans—the hairy strands became longer and then started to branch, first into a few simple tufts and then later into a much more orderly system of barbs projecting sideways from a central shaft. Thus, the quill was born (or, in scientist-speak, the pennaceous feather). Lined up and layered across each other on the arms, these more complex feathers formed wings. Many theropods, particularly paravians, had wings of varying shapes and sizes. Some, like the dromaeosaurid *Microraptor*—one of the first feathered dinosaurs that Xu named and described—even had wings on both the arms and the legs, something unheard of in today's birds.

Wings, of course, are essential for flight. They are the airfoils that provide lift and thrust. For this reason, it was long assumed that wings must have evolved specifically for flight, that some maniraptorans turned their primitive dino-fuzz into sheets of pennaceous quill pens because they were fine-tuning their bodies into airplanes. It's an intuitive explanation, but it's probably false.

In 2008, a team of Canadian researchers was prospecting the badlands of southern Alberta, an area rich in fossils of tyrannosaurs, ceratopsians, duckbills, and other last-surviving Late Cretaceous dinosaurs of North America. At the helm was another polite, even-tempered scientist: Darla Zelenitsky, one of the world's experts on dinosaur eggs and reproduction. Her crew had found the skeleton of a horse-size ornithomimosaur—a beaked, omnivorous ostrich-mimic theropod—and its body was surrounded by wispy dark streaks, some of which seemed to continue straight onto the bone. If they were in Liaoning, Darla told her team with a snide laugh, they could call these things feathers

Darla Zelenitsky collecting dinosaurs in Mongolia.

and announce the find as a career-defining discovery. But they *couldn't* be feathers. This ornithomimosaur was entombed in a sandstone dumped by a river, not rapidly buried in mint condition by Liaoning-style volcanic surges. Plus, no feathered dinosaurs had ever been reported from North America before.

The joke ran its course a year later, when Darla and her team—which also included her husband François Therrien, an expert on dinosaur ecology—found a nearly identical fossil. Another ornithomimosaur, in sandstone, with a mange of cotton-candy fuzz around it. Something strange was going on, so the duo went into the storehouse of the Royal Tyrrell Museum of Palaeontology, where François is a curator, to check out other ornithomimosaurs in the collection. There they found a third fluffy skeleton that had been discovered in 1995—a year before Phil Currie took that photograph of the first feathered theropod from Liaoning and showed it to John Ostrom. The paleontologists who excavated the Albertan fossil in the mid-1990s didn't yet know that dinosaur feathers could be preserved, but Darla and François could tell that the tufts on the three ornithomimosaurs were nearly identical in size, shape, structure, and position to the feathers on many of the Liaoning theropods. This could mean only one thing: they *had* found the first feathered dinosaurs in North America.

The ornithomimosaurs that Darla and François discovered didn't merely have feathers. They also had wings. You can clearly see the black splotches on the arm bones where large, quill-pen-style feathers were anchored, a tidy series of dots and dashes arrayed in lines all up and down the forearm. There's no way this dinosaur could fly, though—it was far too big and heavy, and its arms were far too short and its wings too small to provide a large enough surface area to support the animal in the air. Moreover, it didn't have the huge chest muscles necessary to power flight (the breast muscles of today's birds, whose massive sizes make them good eating) nor the asymmetrical feathers (with a leading vane shorter and stiffer than the trailing vane) that are necessary to withstand the severe forces of surging through an airstream. The same turns out to be true of many of the winged theropods of Liaoning, including *Zhenyuanlong*. They had wings, sure, but their hefty bodies, pathetically undersize wings, and puny frames made them wholly unsuitable for the air.

But why else would a dinosaur develop wings? It may seem like a conundrum, but we have to remember that today's birds use their wings for many things other than flying (which is why, for instance, flightless birds like ostriches don't lose their arms entirely). They are also used as display structures to entice mates and frighten rivals, as stabilizers that help birds climb, as fins to help them swim, and as blankets for keeping eggs warm in the nest, along with many other functions. Wings could have evolved for any of these reasons—or maybe another function entirely—but display seems the most likely, and there is growing evidence for it.

When I was doing my PhD with Mark Norell in New York, there was another student working on his degree a couple of hours north at Yale, in the same department that Ostrom taught in before his death in 2005. Jakob Vinther comes from Denmark, and he has the Viking physique to prove it; he's tall, with sandy blond hair, a big bushy beard, and intense Nordic eyes. Jakob never intended to study dinosaurs—he yearns for the Cambrian Period, that time a few hundred million years before the dinosaurs when life in the oceans was undergoing its big bang. While studying these ancient animals, Jakob started to wonder about how fossil preservation works on the microscopic scale. He began to look at lots of different fossils under high-powered microscopes and realized that many of them preserved a variety of small, bubblelike structures. Comparisons to modern animal tissues showed these to be melanosomes: pigment-bearing vessels. Because melanosomes of different size and shape correspond to different colors—sausage-shaped ones make black; meatball-shaped ones, a rusty red; and so on—Jakob gathered that by looking at fossilized melanosomes, you could tell what colors prehistoric animals would have been when they were alive. We were always told this was impossible, but Jakob proved the experts wrong. In my mind, it's one of the cleverest things a paleontologist has ever done in my lifetime.

Naturally, Jakob decided to take a gander at the newly discovered feathered dinosaurs. If the feathers were preserved well enough, he hoped, they might contain melanosomes. One by one, Jakob and his colleagues in China put the Liaoning dinosaurs under the microscope, and his hunch was proven correct. They found melanosomes everywhere—of all shapes and sizes, orientations, and distributions—which reveal that the feathers of nonflying, winged dinosaurs were a rainbow of different colors. Some were even iridescent, like those of today's shiny-sheened crows. Colorful wings like these would have been perfect display instruments—just like the fabulous tail of a peacock. Although it doesn't definitively prove that these dinosaurs were using their wings for display, it is solid circumstantial evidence.

The totality of the evidence—that wings first evolved in dinosaurs too large and ungainly to fly, that these wings were ornately colored, and that modern birds use their wings for display—has led to a radical new hypothesis. Wings originally evolved as display structures—as advertising billboards projecting from the arms, and in some cases, like *Microraptor*, the legs, and even the tail. Then these fashionably winged dinosaurs would have found themselves with big broad surfaces that by the unbreakable laws of physics could produce lift and drag and thrust. The earliest winged dinosaurs, like the horse-size ornithomimosaurs and even most raptors like *Zhenyuanlong*, probably would have considered the lift and drag produced by their billboards to be little more than an annoyance. In any case, whatever lift was generated wasn't nearly enough to get such large animals into the air. But in more advanced paravians, which had the magic combination of bigger wings and smaller body size, the billboards would have been able to take on an aerodynamic function. These dinosaurs could now move around in the air, even if awkwardly at first. Flight had evolved—and it had happened totally by accident, the billboards now repurposed as airfoils.

The more fossils we find—particularly in Liaoning—the more complex the story gets. The early development of flight appears to have been chaotic. There was no orderly progression, no long evolutionary march in which one subgroup of dinosaurs was refined into ever better aeronauts. Instead, evolution had produced a general type of dinosaur—small, feathered, winged, fast-growing, efficient-breathing—that had all of the attributes needed to start playing around in the air. There seems to have been a zone on the dinosaur family tree where this type of animal had free reign to experiment. Flight probably evolved many times in parallel, as different species of these dinosaurs—with their different airfoil and feather arrangements—found themselves generating lift from their wings as they leaped from the ground, scurried up trees, or jumped between branches.

Some of them were gliders, able only to soar passively on air currents. *Microraptor* undoubtedly could glide, as its arm and leg wings were big enough to support its body in the air. This isn't just conjecture but has been demonstrated by experiments in which scientists have built anatomically correct, life-size models and stuck them in wind tunnels. Not only do they submissively stay afloat, but they're pretty *good* at coasting in the airflow. There's also another type of dinosaur that could probably glide, but in a much different way than *Microraptor*. The tiny *Yi qi*—

maybe the wackiest dinosaur ever found—had a wing, but not made of feathers. Instead, it had a membrane of skin stretching between its fingers and body, like a bat. This membrane must have been a flight structure, but it was not flexible enough to actively flap, so gliding is really the only possibility. The fact that *Microraptor* and *Yi* have such divergent wing configurations is some of the strongest evidence that different dinosaurs were evolving distinct flight styles independently of one another.

Other feathered dinosaurs would have started flying in a different way—by flapping. This is called powered flight, because the animal actively generates lift and thrust by beating its wings. Mathematical models suggest that some non-bird dinosaurs were plausible flappers, including *Microraptor* and the troodontid *Anchiornis*, as both had wings big enough and a body light enough that flapping could have powered them through the air, at least theoretically. These first attempts probably would have been awkward, as these dinosaurs wouldn't have had the muscle strength or stamina to stay in the sky very long, but they provided evolution with a starting point. Now, with these big-winged, small-bodied dinosaurs fluttering around, natural selection could get to work and modify these creatures into better fliers.

One of these wing-beating lineages—maybe the descendants of *Microraptor* or *Anchiornis*, or one that evolved completely separately—got even smaller, developed bigger chest muscles and hyperelongated arms. They lost their tails and teeth, ditched one of their ovaries, and hollowed out their bones even more to lessen their weight. Their breathing became more efficient, their growth faster, and their metabolism more supercharged, so that they became fully warm-blooded, able to maintain a constant high internal body temperature. With each evolutionary enhancement, they became even better fliers, some able to stay airborne for hours on end, others able to sail through the oxygen-starved upper reaches of the troposphere, over the rising Himalayas.

These were the dinosaurs that became the birds of today.

EVOLUTION MADE BIRDS from dinosaurs. And as we've seen, it happened slowly, as one lineage of theropod dinosaurs acquired the characteristic features and behaviors of today's birds piecemeal, over tens of millions of years. A *T. rex* didn't just mutate into a chicken one day, but rather, the transition was so gradual that dinosaurs and birds just seem to blend into each other on the family tree. *Velociraptor*, *Deinonychus*, and *Zhenyuanlong* are on that "non-bird" side of the genealogy, but were they around today, we would probably consider them just another type of bird, no stranger than a turkey or an ostrich. They had feathers, they had wings, they guarded their nests and cared for their babies, and hell, some of them could probably even fly a little bit.

During the tens of millions of years that dinosaurs were evolving the signature features of birds one by one, there was no long game, no greater aim. There was no force guiding evolution to make these dinosaurs ever more adapted to the skies. Evolution works only in the moment, naturally selecting features and behaviors that make an animal successful in its particular time and place. Flight was something that just kind of happened when the time was right. It may have even gotten to the point where it was inevitable. If evolution manufactures a small, long-armed, big-brained hunter with feathers to keep warm and wings to woo mates, it doesn't take very much for that animal to start flapping around in the air. In that moment, working with a fluttering dinosaur with some awkward aerial ability struggling to survive in a dinosaur-eat-dinosaur world, natural selection could kick in and start shaping its progeny into even better fliers. With each additional refinement, you would have something that could fly better, farther, faster—until a modern-style bird had emerged.

The culmination of this long transition was a game-changer in the history of life. When evolution had finally succeeded in assembling a small, winged, flying dinosaur, a great new potential was unlocked. These first birds began to diversify like crazy, probably because they had evolved a new ability that allowed them to invade novel habitats and live a different lifestyle than their predecessors. We can see this (relatively) sudden change in the fossil record.

As part of my PhD project, I joined forces with two number crunchers to assess how rates of evolution changed across the dinosaur-bird transition. Graeme Lloyd and Steve Wang are paleontologists, but I don't know if either of them has ever collected a fossil. They are first-rate statisticians—math whizzes who take joy in sitting in front of their computers for hours, writing code and running analyses.

The three of us worked together to devise a new way of calculating how fast or slow animals change features of their skeletons over time and how these rates change branch by branch on the family tree. We started with the big new genealogy of birds and their closest theropod cousins that I produced with Mark Norell. We then made a vast database of anatomical features that vary in these animals—some species, for example, have teeth, but others, a beak. By mapping the distribution of these characteristics onto the family tree, we could see where one condition changed into another, where teeth gave way to beaks, and so on. This allowed us to count up how many changes occurred on each branch of the tree. We could also figure out how much time each branch on the tree represented, by using the ages of each fossil. Change over time is rate, and thus we could measure the pace of evolution for each branch. Then, using Graeme and Steve's statistical know-how, we could test whether certain time intervals in dinosaur-bird evolution, or certain groups on the family tree, had higher rates of change than others.

The results were about as clear as anything I've ever seen spit out of a statistics software program: most theropods were evolving at ho-hum background rates, but then, once an airworthy bird had emerged, the rates went into overdrive. The first birds were evolving much faster than their dinosaur ancestors and cousins, and they maintained these accelerated rates for many tens of millions of years. Meanwhile, other studies have shown that there was a sudden decrease in body size and a spike in rates of limb evolution right around this same point on the genealogy, as these first birds were quickly getting smaller and growing longer arms and bigger wings so that they could fly better. Although it had taken tens of millions of years for evolution to make a flying bird out of a dinosaur, now things were happening very fast, and birds were soaring.

A QUICK WALK from Xu Xing's office in Beijing is another room, brighter and less solemn but with fewer fossils. It's where Jingmai O'Connor works—but only part of the time. The reason there aren't many fossils here is because Jingmai studies Liaoning birds—the bona fide fliers that flapped over the heads of the feathered dinosaurs—and most of these are crushed onto limestone slabs, so she can describe and measure them from photographs blown up on her computer screen. That means she can easily work from home, which is deep among the last remaining Beijing hutongs—traditional narrow-alleyed neighborhoods of single-story stone buildings pasted together. Good thing too, because she spends a lot of her nonscience time hanging about in the hutongs, raving and even occasionally DJ-ing in the trendy clubs of China's suddenly hip capital.

Jingmai calls herself a paleontologista—fitting, given her fashionista style of leopard-print Lycra, piercings, and tattoos, all of which are at home in the club but stand out (in a good way) among the plaid-and-beard crowd that dominates academia. A native of Southern California—half Irish, half Chinese by blood—Jingmai is a Roman candle of energy—delivering caustic one-liners one moment, speaking in eloquent paragraphs about politics the next, and then it's on to music or art or her own unique personal brand of Buddhist philosophy. Oh yes, and she's also the world's number one expert on those first birds that broke the bounds of Earth to fly above their dinosaur ancestors.

Many birds lived during the Age of Dinosaurs. The first flapping fliers must have originated sometime before 150 million years ago, because that is the age of *Archaeopteryx*, Huxley's Frankenstein creature, which is still, as far as we know, the very oldest true bird, unarguably capable of powered flight, in the fossil record. Most likely, evolution had already assembled a small, winged, flapping, bona fide bird sometime in the middle part of the Jurassic Period, around 170 to 160 million years

LEFT *Yanornis*, a species of true bird—which could fly by flapping its large feathered wings—from Liaoning, China. RIGHT Jingmai O'Connor, the world's leading expert on the oldest bird fossils.

ago. That means there was a good hundred million years during which birds coexisted with their dinosaur predecessors.

A hundred million years is a lot of time to attain a lot of diversity, particularly as these early birds were evolving at such fast rates compared to other dinosaurs. The Liaoning birds that Jingmai studies are a snapshot of this Mesozoic aviary—the best portrait of what birds were doing during the earliest years of their evolutionary history. Every week, middlemen and museum curators from all over China send photographs to Jingmai and her colleagues in Beijing, of new bird fossils plucked from the rolling fields of northeastern China by farmers. Thousands of these fossils have been reported over the last two decades, and they are much more common than feathered dinosaurs like *Microraptor* and *Zhenyuanlong*. It's probably because flocks of primitive birds were suffocated by noxious gases from the big volcanic eruptions, and their limp bodies then fell into the lakes and forests that were buried by the ashy sludge that also entombed the feathered dinosaurs.

Week after week, Jingmai opens her e-mail, downloads the photographs, and finds herself staring at a new type of bird.

These birds include countless species; Jingmai seems to be naming a new one every month or two. They lived in the trees, on the ground, and even in and around the water like ducks. Some of these still had teeth and long tails, retained from their *Velociraptor*-like forebears, whereas others had the tiny bodies, huge breast muscles, stubby tails, and majestic wings of modern birds. Meanwhile, gliding and gawkily flapping alongside these birds were some of those other dinosaurs experimenting with flight—the four-winged *Microraptor*, the bat-winged species, and so on.

This is more or less where things stood 66 million years ago. This whole suite of birds and other airborne dinosaurs was there, gliding and flapping overhead, when *T. rex* and *Triceratops* were duking it out in North America, carcharodontosaurs were chasing titanosaurs south of the equator, and dwarf dinosaurs were hopping across the islands of Europe. And then they witnessed what came next, the instant that snuffed out almost all of the dinosaurs, all but a few of the most advanced, best-adapted, best-flying birds, which made it through the carnage and are still with us today—among them the seagulls outside my window.

9

DINOSAURS DIE OUT

Edmontosaurus

IT WAS THE WORST DAY in the history of our planet. A few hours of unimaginable violence that undid more than 150 million years of evolution and set life on a new course.

T. rex was there to witness it.

When a pack of Rexes woke up that morning 66 million years ago on what would go down as the final day of the Cretaceous Period, all seemed normal in their Hell Creek kingdom, the same as it had for generations, for millions of years.

Forests of conifers and ginkgos stretched to the horizon, interspersed with the bright flowers of palms and magnolias. The distant churn of a river, rushing eastward to empty into the great seaway that lapped against western North America, was drowned out by the low bellow of a herd of *Triceratops* several thousand strong.

As the pack of *T. rex* readied themselves for the hunt, sunlight began to trickle through the forest canopy. It highlighted the outlines of various small critters darting through the sky, some flapping their feathered wings and others gliding on currents of hot air rising from the humidity of the young day. Their chirps and tweets were beautiful, a dawn symphony that could be heard by all the other creatures of the forest and floodplains: armored ankylosaurs and dome-headed pachycephalosaurs hiding in the trees, legions of duckbills just beginning their breakfast of flowers and leaves, raptors chasing mouse-size mammals and lizards through the brush.

Then things started to get weird, truly outside all norms of Earth history.

For the last several weeks, the more perceptive of the Rexes may have noticed a glowing orb in the sky, far off in the distance—a hazy ball with a fiery rim, like a duller and smaller version of the sun. The orb seemed to be getting larger, but then it would disappear from view for large portions of the day. The Rexes wouldn't have known what to make of it; it was far beyond their brainpower to contemplate the motions of the heavens.

But this morning, as the pack broke through the trees and emerged onto the riverbank, all of them could see that something was different. The orb was back, and it was gigantic, its shine illuminating much of the sky to the southeast in a cloudy psychedelic mist.

Then, a flash. No noise, only a split-second flare of yellow that lit up the whole sky, disorienting the Rexes for a moment. As they blinked their eyes back to focus, they noticed that the orb was now gone, the sky a dull blue. The alpha male turned to check on the rest of his pack. . . .

And then they were blindsided. Another flash, but this one far more vengeful. The rays lit the morning air in a fireworks display and burned into their retinas. One of the juvenile males fell over, cracking his ribs. The rest of them stood frozen, blinking manically, trying to rid themselves of the sparks and speckles that flooded their vision. Still no sound to go with the visual fury. In fact, no noise at all. By now, the birds and flying raptors had stopped chirping, and silence hung over Hell Creek.

The calm lasted for only a few seconds. Next, the ground beneath their feet started to rumble, then to shake, and then to flow. Like waves. Pulses of energy were shooting through the rocks and soil, the ground rising and falling, as if a giant snake were slithering underneath. Everything not rooted into the dirt was thrown upward; then it crashed down, and then up and down again, the Earth's surface having turned into a trampoline. Small dinosaurs and the little mammals and lizards were catapulted upward, then splattered onto trees and rocks when they landed. The victims danced across the sky like shooting stars.

Even the largest, heaviest, forty-foot-long Rexes in the pack were launched several feet off the ground. For a few minutes, they bounced around helplessly, flailing about as they rode the trampoline. Moments earlier they had been the undisputed despots of an entire continent; now they were little more than seven-ton pinballs, their limp bodies careening and colliding through the air. The forces were more than enough to crush skulls, snap necks, and break legs. When the shaking finally stopped and the ground was no longer elastic, most of the Rexes were littered along the riverbank, casualties on a battlefield.

Very few of the Rexes—or the other dinosaurs of Hell Creek—were able to walk away from the bloodbath. But some did. As the lucky survivors staggered out, sidestepping the corpses of their compatriots, the sky began to change color above them. Blue turned to orange, then to pale red. The red got sharper and darker. Brighter, brighter, brighter. As if the headlights of a giant oncoming car were coming closer and closer. Soon everything was bathed in an incandescent glory.

Then the rains came. But what fell from the sky was not water. It was beads of glass and chunks of rock, each one scalding hot. The pea-size morsels pelted the surviving dinosaurs, gouging deep burns into their flesh. Many of them were gunned down, and their shredded corpses joined the earthquake victims on the battlefield. Meanwhile, as the bullets of glassy rock whizzed down from above, they were transferring heat to the air. The atmosphere grew hotter, until the surface of the Earth became an oven. Forests spontaneously ignited and wildfires swept across the land. The surviving animals were now roasting, their skin and bones cooking at temperatures that instantaneously produce third-degree burns.

It was no more than fifteen minutes since the *T. rex* pack was startled by that first jolt of light, but by now they were all dead, as were most of the dinosaurs they had lived with. The once-lush woodlands and river valleys were aflame. Still, animals had survived—some of the mammals and lizards were underground, some of the crocodiles and turtles were underwater, and some of the birds had been able to fly to safer refuges.

Over the next hour or so, the rain of bullets ceased, and the air cooled. A breath of calm once again settled over Hell Creek. It seemed that the danger was over, and many of the survivors came out of their hiding places to survey the scene. Carnage everywhere, and although the sky was no longer radioactive red, it was getting blacker as it choked up with soot from the forest fires, which were still raging. As a couple of raptors sniffed the charred bodies of the *T. rex* pack, they must have thought that they had survived the apocalypse.

They were wrong. Some two and a half hours after the first light flash, the clouds began to howl. The soot in the atmosphere began to swirl into tornadoes. And then—*woosh*—the wind charged across the plains and through the river valleys, blowing at hurricane force, hard enough to make many of the rivers and lakes burst their banks. Along with the wind was a deafening noise, louder than anything these dinosaurs had ever heard. Then another. Sound travels much slower than light, and these were the sonic booms that occurred at the same time as the two light flashes, caused by the distant horror that had started the chain reaction of brimstone hours earlier. The raptors shrieked in pain as their ears ruptured, and many of the smaller critters hurried back into the safety of their burrows.

While all of this was happening in western North America, other parts of the world were going through their own upheavals. The earthquakes, glassy-rock rain, and hurricane winds were less severe in South America, where carcharodontosaurs and giant sauropods roamed. The same was true of the European islands that the weird Romanian dwarf dinosaurs called home. Still, these dinosaurs also had to deal with quaking ground, wildfires, and intense heat, and many of them died during those same chaotic two hours that wiped out most of the Hell Creek community. Other places, though, had it much worse. Much of the mid-Atlantic coast was sliced apart by tsunamis twice as tall as the Empire State Building, which flushed the carcasses of plesiosaurs and other sea-dwelling giant reptiles far inland. Volcanoes started to spew out rivers of lava in India. And a zone of Central America and southern North America—everything within a radius of about six hundred miles (one thousand kilometers) of the Yucatán Peninsula of modern-day Mexico—was annihilated. Vaporized.

As the morning gave way to afternoon and then evening, the winds died down. The atmosphere continued to cool, and although there were a few aftershocks, the ground was stable and solid. The wildfires seared away in the background. When night finally came and this most horrible of days finally was over, many—maybe even most—of the dinosaurs were dead, all over the world.

Some did stagger on, however, into the next day, the next week, the next month, the next year, and the next decades. It was not an easy time. For several years after that terrible day, the Earth turned cold and dark because soot and rock dust lingered in the atmosphere and blocked out the sun. The darkness brought cold—a nuclear winter that only the hardiest of animals could survive. The darkness also made it very difficult for plants to subsist, as they need sunlight to power photosynthesis to make their food. As plants died, food chains collapsed like a house of cards, killing off many of the animals that had been able to endure the cold. Something similar happened in the oceans, where the death of photosynthesizing plankton took out the larger plankton and fish that fed on them and in turn the giant reptiles at the top of the food pyramid.

The sun did eventually break through the darkness, as the soot and other gunk was leached out of the atmosphere by rainwater. These rains, however, were highly acidic and would have scalded much of the Earth's surface. And the rain was not able to remove some ten trillion tons of carbon dioxide that had been blown up into the sky with the soot. CO_2 is a nasty greenhouse gas that traps heat in the atmosphere, and soon nuclear winter gave way to global warming. All of these things conspired in a war of attrition to knock off whatever dinosaurs were not felled by the initial cocktail of earthquakes, brimstone, and fires.

A few hundred years after that dreadful day—a few thousand years at the absolute most—western North America was a scarred, post-apocalyptic landscape. What was once a diverse ecosystem of sweeping forests, alive with the hoofbeats of *Triceratops* and ruled by *T. rex*, was now quiet and mostly empty. Here and there, the odd lizard scurried through the bushes, some crocodiles and turtles paddled in the rivers, and rat-size mammals periodically peeked out of their burrows. A few birds were still around, picking at seeds still buried in the soil, but all the other dinosaurs were gone.

Hell Creek had turned to Hell. So had much of the rest of the world. It was the end of the Age of Dinosaurs.

WHAT HAPPENED ON that day—when the Cretaceous ended with a bang and the dinosaurs' death warrant was signed—was a catastrophe of unimaginable scale that, thankfully, humankind has never experienced. A comet or an asteroid—we aren't sure which—collided with the Earth, hitting what is now the Yucatán Peninsula of Mexico. It was about six miles (ten kilometers) wide, or about the size of Mount Everest. It was probably moving at a speed of around 67,000 miles per hour (108,000 kilometers per hour), more than a hundred times faster than a jet airliner. When it slammed into our planet, it hit with the force of over 100 trillion tons of TNT, somewhere in the vicinity of a billion nuclear bombs' worth of energy. It plowed some twenty-five miles (forty kilometers) through the crust and into the mantle, leaving a crater that was over 100 miles (160 kilometers) wide.

The impact made an atom bomb look like a Fourth of July cherry bomb. It was a bad time to be alive.

The Hell Creek dinosaurs were living about 2,200 miles (3,500 kilometers) northwest of ground zero, as the *Microraptor* flies. Give or take a little artistic license, they would have experienced the string of terrors described above. Their cousins in New Mexico—southern versions of *T. rex*, other types of horned and duck-billed dinosaurs, and some of the few sauropods living in North America, whose bones I've collected

Earth forty-five seconds after the impact of the Chicxulub asteroid, with a growing cloud of dust and molten rock shooting into the atmosphere and a fire-igniting heat pulse starting to spread across the oceans and land. *Artwork by Donald E. Davis, NASA.*

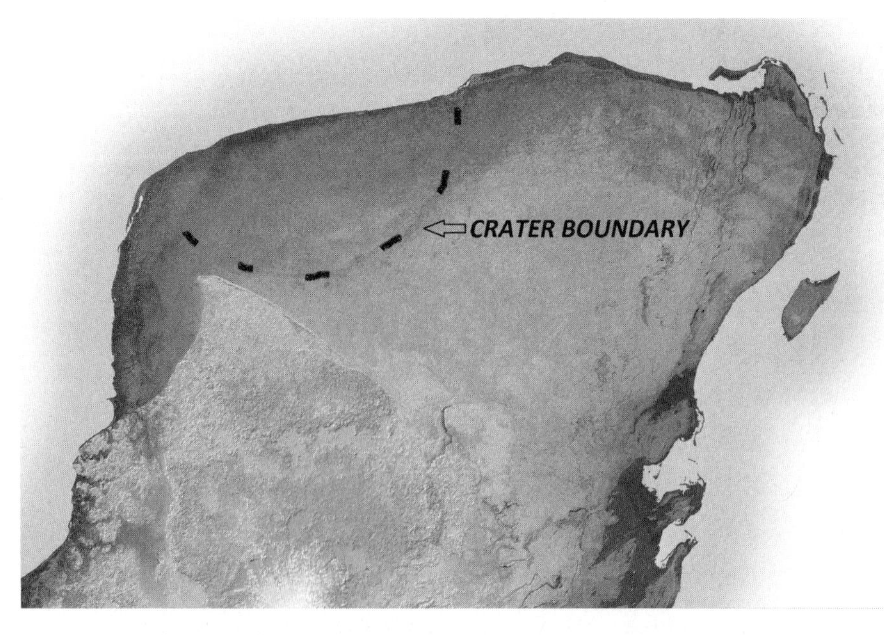

Relief map of the modern-day Yucatán Peninsula of Mexico, showing the outline of the Chicxulub crater (the remainder of the crater is underwater). *Courtesy of NASA.*

during many summers of fieldwork—would have been even worse off. They were only about 1,500 miles (2,400 kilometers) from the impact site. The closer you were, the greater the horrors: the light and sound pulses would have arrived quicker, the earthquakes would have been more severe, the rain of glass and rocks would have been heavier, and the temperature of the oven would have been greater. All creatures living within six hundred miles (a thousand kilometers) or so of the Yucatán would have been instantly turned into ghosts.

The glowing orb in the sky, which piqued the interest of the *T. rex* herd, was the comet or asteroid itself (from here on out, I'll just refer to it as an asteroid for simplicity). If you were around back then, you would have seen it. The experience would have probably been similar to those times Halley's Comet has come close to Earth. Seemingly floating up in the heavens, the asteroid would have appeared harmless. You would have been oblivious, at least at first.

The first flash of light occurred as the asteroid punched through the Earth's atmosphere and violently compressed the air in front of it, so much that the air became four or five times as hot as the surface of the sun and ignited. The second flash was the impact itself, when asteroid met bedrock. The sonic booms associated with both of these flashes followed many hours later, sound moving much slower than light. With them came the winds, which probably blasted at over 600 miles per hour (1,000 kilometers per hour) close to the Yucatán and still at several hundred miles per hour by the time they reached Hell Creek. (For comparison, Hurricane Katrina's maximum wind speed was measured at about 175 miles per hour.)

As the asteroid and Earth smashed together, an enormous

amount of energy was unleashed, which fed shock waves that caused the ground to shake like a trampoline. These earthquakes were probably around 10 on the Richter scale—far more powerful than anything human civilizations have ever coped with. Some of these earthquakes triggered the Atlantic tsunamis, which ripped up house-size boulders and flung them far inland; others kicked the Indian volcanoes into hyperdrive, and they kept erupting for thousands of years, compounding everything else the asteroid had wrought.

The energy from the collision vaporized the asteroid and the bedrock that it hit. Dust, dirt, rock, and other debris from the collision shot up into the sky—most as vapor or liquid but some as small but still solid pieces of rock. Some of this material flew past the outer fringes of the atmosphere into outer space. But what goes up (as long as it doesn't reach escape velocity) must come down, and as it did, the liquified rock cooled into glassy blobs and tear-shaped spears, which transferred heat to the atmosphere, transforming it into an oven.

The spiking temperatures lit forests on fire—maybe not all over the world, but certainly in much of North America and anywhere else within a few thousand miles of the Yucatán. We see the singed remnants of leaves and wood—the kind of stuff left after a campfire has been extinguished—in rocks that were laid down right after the asteroid hit. The soot from the fires, with other dust and grime kicked up by the impact but too light to fall back down to earth, would have floated up into the atmosphere, clogging the currents that circulate air across the globe, until the entire planet was dark. The ensuing period—thought to be equivalent to a global nuclear winter—probably killed off most of the dinosaurs in areas far from the smoldering crater.

I could go on and on, exhausting my thesaurus, but if I go much further, you probably won't believe me. Which would be a shame, because all that I write really happened. And we know that because of the work of one man, a geological genius who is one of my scientific heroes: Walter Alvarez.

WE'VE ALREADY ESTABLISHED that I did some silly things in high school, when my obsession with dinosaurs overtook my better judgment. My fanboy stalking of Paul Sereno wasn't nearly the worst of it. Nothing was more brazen than when I picked up the phone one day in the spring of 1999 and cold-called Walter Alvarez at his office in Berkeley, California. I was a fifteen-year-old kid with a rock collection; he was the eminent National Academy of Sciences member who nearly twenty years earlier had proposed the idea that a giant asteroid impact killed off the dinosaurs.

He answered on the second ring. Even more astounding, he didn't hang up as I rambled on about the purpose of my call. I had read his book *T. rex and the Crater of Doom*—still, to my mind, one of the best pop-science books on paleontology ever written—and was captivated by how he put together the clues that pointed to the asteroid. His book explained how the detective game started in a rocky gorge on the outskirts of the medieval commune of Gubbio, in the Apennine Mountains of Italy. It was here where Alvarez first noticed the unusual character of the thin band of clay that marked the end of the Cretaceous. As chance had it, my family was gearing up for a trip to Italy to celebrate my parents' twentieth wedding anniversary. It would be my first time outside of North America, and I wanted to make it memorable. For me, that wasn't basilicas and art museums, but a pilgrimage to Gubbio, to stand on the spot where Alvarez started to figure out one of the biggest riddles in science.

But I needed directions, so I decided to go straight to the source.

Professor Alvarez gave me detailed instruction that even a kid without any modicum of Italian could follow. We also talked for a while about my interest in science. Looking back, I am astonished that such a scientific giant could be as kind and generous with his time as he was. But alas, it turned out to be for naught, because my family never made it to Gubbio that summer. Floods closed the main rail line from Rome, and I was devastated. My whining nearly ruined my parents' second honeymoon.

Five years later, however, I was back in Italy for a college geology field course. We were staying in a small observatory in the Apennines run by Alessandro Montanari, one of the many scientists who made a name in the 1980s studying the end-Cretaceous extinction. On our first-day tour we passed through the library, where a solitary figure was scrutinizing a geological map under a flickering light.

"I want you all to meet my friend and mentor, Walter Alvarez," Sandro said in his singsong Italian accent. "Some of you may have heard of him."

I was paralyzed. Never, before or since, have I been as gobsmacked. The rest of the tour was a blur, but afterward I sneaked back to the library and gently opened the door. Alvarez was still there, hunched over the map in a trance of concentration. I felt bad about interrupting him—maybe he was homing in on some other unsolved mystery of Earth history. I introduced myself and was gobsmacked a second time when he remembered our conversations from a few years back.

"Did you make it to Gubbio back then?" he asked me.

I could only mutter an embarrassed no, not really wanting to admit that I had wasted his time with that phone call—and several e-mails that followed.

"Well then, get ready, because I'm taking your class there in a few days," he replied. I flashed a megawatt smile.

Days later we were in Gubbio, gathered in the gorge, Mediterranean sun beaming down and fast cars whizzing by, a fourteenth-century aqueduct perched precariously on the cliffs above. Walter Alvarez stepped in front of us. His khakis were stuffed with rock samples; he wore a wide-brimmed hat and reflective aquamarine shirt to ward off the sun. He pulled his hammer out of its holster, and pointed downward and to his right, to a thin gouge in the rock that cut through the rosy pink limestone forming most of the gorge. This rock was softer, finer; it was a layer of clay, about one centimeter thick, a bookmark separating the limestones of the Cretaceous below from those of the postextinction Paleogene period above. It was here—this man, standing at this spot, looking at this strip of clay—where the asteroid theory was conceived a quarter century earlier.

Afterward, we stopped for truffle pasta, white wine, and biscotti at a five-hundred-year-old restaurant just down the road. Before our lunch, we dutifully signed a leather guest book, inscribed with the names of many of the geologists and paleontologists who have come through Gubbio to study the gorge and its celebrity clay. It read like a Hall of Fame roster, and I've never taken more pride in signing my name. For the next two

Walter Alvarez pointing to the boundary between the Cretaceous rocks (below) and Paleocene rocks (above) in Gubbio, Italy. The boundary is the divot located between his rock hammer and right knee. *Courtesy of Nicole Lunning.*

hours, I sat across from Walter as, between mouthfuls of linguine, he told my starstruck classmates and me the story of how he cracked the dinosaur mystery.

In the early 1970s, not long after Walter finished his PhD, the plate-tectonics revolution had consumed the science of geology, and people now realized that continents moved around over time. One way you could track their motions was by looking at the orientation of small crystals of magnetic minerals, which point themselves toward the North Pole when lavas or sediments harden into stone. Walter reckoned that this new science of paleomagnetism could help untangle how the Mediterranean region was assembled—how small plates of crust rotated and crushed into each other to form modern-day Italy and raise the Alps. That is what first brought him to Gubbio, to measure microscopic bits of minerals within the thick limestone sequence of the gorge. But when he was there, he became intrigued by an even bigger mystery. Some of the rocks he was measuring were crammed with fossil shells of all shapes and sizes, which belonged to a great diversity of creatures called forams—tiny predators that float around in the ocean plankton. Above these rocks, however, were nearly barren limestones, sprinkled with a few tiny, simple-looking forams.

Walter was observing a line between life and death. It's the geological equivalent of listening to those last few moments on a cockpit voice recorder before it gives way to silence.

Walter wasn't the first person to notice it. Geologists had been working in the gorge for decades, and painstaking work by an Italian student named Isabella Premoli Silva had determined that the diverse forams were Cretaceous in age, the simple ones from the Paleogene. The knife-edge separation between them corresponded to what had long been recognized as a mass extinction—one of those unusual times in Earth history when lots of species disappear simultaneously all over the world.

But this wasn't your average mass extinction. Specks of plankton weren't its only casualties, and it wasn't confined to the water. It decimated the oceans and the land, and killed off many other types of plants and animals.

Including the dinosaurs.

No way could that be coincidence, Walter thought. What happened to the forams must have been linked to what happened to the dinosaurs and all of the other things that perished, and he wanted to figure it out.

The key, he realized, was hidden in that tiny strip of clay between the fossil-rich Cretaceous limestones and the sterile Paleogene ones. But when he first saw it, it didn't seem all that special. It wasn't heaving with mangled fossils, streaked with flamboyant colors, or rotten in scent. It was just clay, so fine that you couldn't even see the individual grains with the naked eye.

Walter called his dad for help. His father just so happened to be a Nobel Prize–winning physicist: Luis Alvarez, who had discovered a host of subatomic particles and had been one of the key players in the Manhattan Project. (He even flew behind the *Enola Gay* to monitor the effects of Little Boy when it was dropped on Hiroshima.) Alvarez the younger thought that Alvarez the elder might have some unconventional ideas for chemically analyzing the clay. Maybe there was something hidden in there that could tell them how long it took the thin layer to form. If it formed gradually, the product of millions of years of slow accumulation of dust in the deep ocean, then the death of the forams, and thus of the dinosaurs, was a drawn-out affair.

But if it was deposited suddenly, that meant the Cretaceous must have ended in catastrophe.

Measuring the length of time it took a rock layer to form is tricky—one of the headaches faced by all geologists. But in this case, the father-and-son team came up with what they figured was a clever solution. Heavy metals—some of those elements in the nether regions of the periodic table, like iridium—are rare on Earth's surface, which is why most people have never heard of them. But tiny amounts of them fall at a more or less constant rate from the deep reaches of outer space as cosmic dust. The Alvarezes reasoned that if the clay layer had only a tiny peppering of iridium, then it had formed very quickly; if it had a larger amount, then it must have formed over a much longer time period. New instruments now allowed scientists to measure even very small concentrations of iridium, including one in a lab at Berkeley run by one of Luis Alvarez's colleagues.

They weren't prepared for what they found.

They found iridium all right—lots of it. Too much of it. There was so much iridium that it would have taken many tens of millions of years—maybe even hundreds of millions of years—of steady cosmic dusting to deliver it all. Which was impossible, because the limestones above and below the clay were dated well enough that the Alvarezes knew that the clay layer could have been deposited over only a few million years at most. Something was amiss.

Maybe it was a mistake, some local quirk of the Gubbio gorge. So they went to Denmark, where rocks of the same age jut into the Baltic Sea. Here, too, they found an iridium anomaly right at the Cretaceous-Paleogene boundary. Before long, a tall young Dutchman named Jan Smit caught wind of what the Alvarezes were doing and reported that he had also been sniffing around for iridium—and had found a spike at the boundary in Spain. More reports of iridium soon followed, from rocks formed on land, in shallow water, and in the deep ocean, all at that fateful moment when dinosaurs disappeared.

The iridium anomaly was real. The Alvarezes went through the possible scenarios: volcanoes, flooding, climate change, and others, but only one made sense. Iridium is super-rare on Earth but much more common in outer space. Could something from the deep expanses of the solar system have delivered an iridium bomb 66 million years ago? Perhaps it was a supernova explosion, but more likely a comet or an asteroid. After all, as the many craters pockmarking the surface of the Earth and moon attest to, these interstellar visitors do occasionally bombard us. It was a bold idea, but not a crazy one.

Luis and Walter Alvarez, with their Berkeley colleagues Frank Asaro and Helen Michel, published their provocative theory in *Science* in 1980. It unleashed a decade of scientific frenzy. Dinosaurs and mass extinctions were constantly in the news, the impact hypothesis was debated in countless books and television documentaries, a dinosaur-killing asteroid made the cover of *Time*, and hundreds of scientific papers went back and forth on what really killed the dinosaurs, with scientists as diverse as paleontologists, geologists, chemists, ecologists, and astronomers weighing in on the hottest scientific issue of the day. There were feuds, egos clashed, but the crucible of fierce debate put everyone at the top of their game, as they gathered (or disputed) evidence for an impact.

By the end of the 1980s, it was undeniable that the Alvarezes were correct: an asteroid or comet did hit the planet 66 million years ago. Not only was the same iridium layer found all over the world, but other geological oddities pointing to an impact were found alongside the iridium. There was a strange type of quartz in which the mineral planes had collapsed, leaving a telltale sign of parallel bands shooting through the crystal structure. This "shocked quartz" had previously been found in only two places: the rubble of nuclear bomb tests and the inside of meteor craters, formed from the fierce shock waves of these explosive events. There were spherules and tektites—spherical or spear-shaped bullets of glass forged from the melted products of a big collision that cooled as they fell back down through the atmosphere. Tsunami deposits were discovered around the Gulf of Mexico, dating right to the Cretaceous-Paleogene boundary, showing that a monumental event caused monstrous earthquakes right when the quartz was being shocked and the tektites were falling.

Then, as the 1990s dawned, the crater was finally found. The smoking gun. It had taken a while to find it because it was buried under millions of years of sediment in the Yucatán. The only detailed studies of the area had been carried out by oil-company geologists who kept their maps and samples locked up for many years. But there could be no doubt: the 110-mile-wide (180 km) hole buried under Mexico, called the Chicxulub crater, was dated right to the end of the Cretaceous, 66 million years ago. It is one of the largest craters on Earth, a sign of just how big the asteroid was, how catastrophic the impact. It was probably one of the biggest, perhaps *the* biggest asteroid to hit Earth in the last half billion years. The dinosaurs probably didn't stand a chance.

BIG DEBATES IN science—particularly those that spill out of the specialist journals and into the public eye—always attract skeptics. So it was with the asteroid theory. These dissenters couldn't argue that there was no asteroid—the discovery of the Chicxulub crater made such a claim foolish. Instead, they contended that the asteroid was wrongly accused, an innocent bystander that just so happened to smash into the Yucatán when dinosaurs and the many other things that died out at the end of the Cretaceous—the flying pterosaurs and sea-living reptiles, the coiled ammonites, the big and diverse foram communities in the ocean, and many others—were already on their way out. At worst, the asteroid was the coup de grace that finished a holocaust nature had already started.

It might seem too coincidental to take seriously—a six-mile-wide asteroid arriving exactly when thousands of species were already on their deathbed. However, unlike the flat-earthers and global-warming deniers, these skeptics had a point. When the asteroid fell from the sky, it didn't rudely interrupt some kind of static, idyllic, lost world of the dinosaurs. No, it hit a planet that was in quite a bit of chaos. The big volcanoes in India that the asteroid kicked into overdrive had actually started erupting a few million years before. Temperatures were gradually getting cooler, and sea levels were fluctuating dramatically. Maybe some of these things factored into the extinction? Perhaps they were the primary culprits; maybe these longer-term environmental changes were causing dinosaurs to slowly waste away.

The only way to test these ideas against each other is to look very closely at the evidence that we have—dinosaur fossils. What we have to do is track dinosaur evolution over time, to see if there are any long-term trends and see what changes occurred at or near the Cretaceous-Paleogene boundary, when the asteroid hit. This is where I enter the picture. From the time I first spoke with Walter Alvarez on the phone, I was hooked on the riddle of the dinosaur extinction. My addiction spiraled as I stood next to Walter in the Gubbio gorge. Then, as a graduate student I finally had a chance to make my own contribution to the debate, using one of the specialties I had developed as a young researcher: using big databases and statistics to study evolutionary trends.

My venture into the extinction debate was a joint one, with my old friend Richard Butler. A few years earlier, we were bushwhacking through Polish quarries on the hunt for footprints of the very oldest dinosaurs; now in 2012 as I was starting to wrap up my PhD, we wanted to know why the descendants of these wispy ancestors disappeared over 150 million years later, after they became so phenomenally successful. The question we asked ourselves was this: how were dinosaurs changing during the 10 to 15 million years before the asteroid hit? The way we addressed it was using morphological disparity, the same metric that I used to study the very oldest dinosaurs, which quantifies the amount of anatomical diversity over time. Increasing or stable disparity during the latest Cretaceous would indicate that dinosaurs were doing rather well when the asteroid came, whereas declining disparity would suggest they were in trouble and maybe already on their way to extinction.

We crunched the numbers and found some intriguing results. Most dinosaurs had relatively steady disparity during that last gasp before the impact, including the meat-eating theropods, long-necked sauropods, and small to midsize plant-eaters like the dome-skulled pachycephalosaurs. There was no sign that anything was wrong with them. But two subgroups were in the midst of a disparity decline: the horned ceratopsians like *Triceratops* and the duck-billed dinosaurs. These were the two main groups of large-bodied plant-eaters, which consumed enormous amounts of vegetation with their sophisticated chewing and leaf-shearing abilities. If you were standing around during the latest Cretaceous—anytime between about 80 and 66 million years ago—it was these dinosaurs that would have been most abundant, at least in North America where the fossil record of this time is best. They were the cows of the Cretaceous, the keystone herbivores at the base of the food chain.

Around the same time we were doing our study, other researchers were examining the dinosaur extinction from other angles. Teams led by Paul Upchurch and Paul Barrett in London undertook a census of dinosaur species diversity over the course of the Mesozoic—a simple count of how many dinosaurs were alive at every given point of their reign, corrected for biases caused by the uneven quality of the fossil record. They found that dinosaurs as a whole were still very diverse at the time the asteroid hit, as numerous species were frolicking throughout not only North America but the entire planet. Curiously enough, however, the horned and duck-billed dinosaurs underwent a decline in species numbers right at the end of the Cretaceous, coincident with their decline in disparity.

What would all of this have meant in real-world terms? After all, it was a curious mix: most dinosaurs doing fine, but the big plant-guzzlers showing signs of stress. This question was addressed by a clever computer modeling study by one of the new breed of highly quantitative graduate students: Jonathan Mitchell from the University of Chicago. Jon and his team built food webs for several Cretaceous dinosaur ecosystems, based on careful review of all the fossils that had been found at particular field sites—not only the dinosaurs, but everything they lived with, from crocodiles and mammals down to insects. Then they used computers to simulate what would happen if a few species were knocked out. The result was startling: those food webs that existed when the asteroid struck, which had fewer large herbivores at their bases because of the diversity decline, collapsed easier than the more diverse food webs from only a few million years before the impact. In other words, the loss of some of the big herbivores, even without the decline of any of the other dinosaurs, made end-Cretaceous ecosystems highly vulnerable.

Statistical analyses and computer simulations are all well and good, and there's no doubt that they are the future of dinosaur research, but they can be a little abstract, and sometimes it's useful to simplify things. In paleontology, that means going back to the fossils themselves: holding them in your hands and thinking deeply about them as living, breathing animals, considering them as the very animals that first had to cope with those Late Cretaceous volcanic eruptions and temperature and sea-level shifts, then later stare down an asteroid the size of a mountain.

What we really want to study are the fossils of those last surviving dinosaurs, the ones that witnessed or came close to witnessing the asteroid do its dirty work. Unfortunately, there are only a few places in the world that preserve these types of fossils—but they are starting to tell a convincing story.

The most famous place, without doubt, is Hell Creek. People have been collecting the bones of *T. rex*, *Triceratops*, and their contemporaries for well over a hundred years now throughout the upper Great Plains of the American West. The rocks of Hell Creek are very well dated, too. And that means you can track the diversity and abundance of dinosaurs through time, right up to the iridium layer that fingerprints the asteroid. A number of scientists have done just that—my friend David Fastovsky (author of the best dinosaur textbook on the market) and his colleague Peter Sheehan, a team led by Dean Pearson, and other crews led by Tyler Lyson, a gifted young scientist who grew up on a sprawling ranch in North Dakota in the heart of some of the best dinosaur-bone badlands. They've all found the same thing: dinosaurs were thriving all throughout the time the Hell Creek rocks were laid down, as the Indian volcanoes were erupting and temperatures and sea levels were changing, right up to that moment the asteroid hit. There are even *Triceratops* bones a few centimeters below the iridium. It seems that the asteroid caught the residents of Hell Creek blissfully unaware, right at the peak of their glory days.

Things were similar in Spain, where important new discoveries are emerging from the Pyrenees, along the border with France. This area is being scoured by an energetic duo of thirty-something paleontologists—Bernat Vila and Albert Sellés, two of the most dedicated guys I know, who often find themselves working for months on end without a salary, victims of Spain's torturously slow recovery from a series of financial crises that began in the late 2000s. Somehow that hasn't stopped them. They keep finding dinosaur bones, teeth, footprints, and even eggs. These fossils show that a diverse community—including theropods, sauropods, and duckbills—persisted here into the very latest Cretaceous with no indication that anything was amiss. It's interesting that, a few million years before the asteroid hit, there was a brief turnover event, when armored dinosaurs disappeared locally and more primitive plant-eaters were replaced by advanced duckbills. It's possible that this is related to the decline of the big plant-eaters in North America, although this is hard to prove. It may be that changes in sea level were to blame; as seas rose and fell, they carved up the land that dinosaurs could live on, which led to some small changes in the composition of ecosystems.

Finally, the story appears to be the same in Romania, where Mátyás Vremir and Zoltán Csiki-Sava have been collecting a great diversity of latest Cretaceous dinosaurs, and also in Brazil, where Roberto Candeiro and his students keep finding more teeth and bones of big theropods and enormous sauropods that probably made it to the end. The drawbacks of these places are that the rocks are still not dated very well, so we can't be absolutely sure where the dinosaur fossils sit relative to the Cretaceous-Paleogene boundary, but no doubt the dinosaurs in both areas are latest Cretaceous in age, and there are no signs that they were in any type of trouble.

There was so much new evidence from fossils, statistics, and computer modeling that Richard Butler and I figured the time had come to synthesize it. We came up with something of a dangerous idea: perhaps we could recruit a crack team of dinosaur experts to sit down, discuss everything we currently know about the dinosaur extinction, and try to come to a consensus on why we thought dinosaurs died out. Paleontologists had been arguing for decades on this topic, and in fact it was dinosaur workers who were some of the most ardent skeptics of the asteroid hypothesis in the 1980s. We thought our subversive little plot might end in deadlock or, worse, in a shouting match, but quite the opposite happened. Our team came to an agreement.

Dinosaurs were doing well in the latest Cretaceous. Their overall diversity—both in terms of species numbers and anatomical disparity—was fairly stable. It had not been gradually declining for millions and millions of years, nor was it clearly increasing. The major groups of dinosaurs all persisted into the very latest Cretaceous—theropods big and small, sauropods, horned and duck-billed dinosaurs, dome-headed dinosaurs, armored dinosaurs, smaller plant-eaters, and omnivores. At least in North America, where the fossil record is best, we know that *T. rex*, *Triceratops*, and the other Hell Creek dinosaurs were there when the asteroid destroyed much of the Earth. All of these facts rule out the once popular hypothesis that dinosaurs wasted away gradually due to long-term changes in sea level and temperature or that the Indian volcanoes had started to pick away at the dinosaurs earlier in the Late Cretaceous, a few million years before the end.

Instead, we found that there is no doubt about it: the dinosaur extinction was abrupt, in geological terms. This means that it happened over the course of a few thousand years at most. Dinosaurs were prospering, and then they simply disappear from the rocks, simultaneously all over the world, wherever latest Cretaceous rocks are known. We never find their fossils in the Paleogene rocks laid down after the asteroid impact—nothing, not a single bone or a single footprint anywhere. This means a sudden, dramatic, catastrophic event is likely to blame, and the asteroid is the obvious culprit.

However, there is a nuance. The big herbivores did undergo

a bit of a decline right before the end of the Cretaceous, and the European dinosaurs experienced a turnover as well. This decline apparently had consequences: it made ecosystems more susceptible to collapse, making it more likely that the extinction of just a few species would cascade through the food chain.

All told, then, it appears the asteroid came at a horrible time for the dinosaurs. If it had hit a few million years earlier, before the dip in herbivore diversity and perhaps the European turnover, ecosystems would have been more robust and would have been in a better position to deal with the impact. If it happened a few million years later, maybe herbivore diversity would have recovered—as it had countless other times over the preceding 150-plus million years of dinosaur evolution, when small diversity declines occurred and were corrected—and ecosystems again would have been more robust. There's probably never a good time for a six-mile-wide asteroid to shoot down from the cosmos, but for dinosaurs, 66 million years ago may have been among the worst possible times—a narrow window when they were particularly exposed. If it had happened a few million years earlier or later, maybe it wouldn't just be seagulls congregating outside my window but tyrannosaurs and sauropods too.

Or perhaps not. It's possible the massive asteroid would have done them in regardless. Maybe there was no escape from something that big, packing that kind of punch when it barreled its way into the Yucatán. Whatever the exact sequence of events, I'm confident the asteroid was the primary reason that the non-bird dinosaurs died out. If there is one, single straightforward proposition that I would stake my career on, it would be this: no asteroid, no dinosaur extinction.

THERE IS ONE final puzzle that I haven't addressed yet. Why did all the non-bird dinosaurs die at the end of the Cretaceous? After all, the asteroid didn't kill everything. Plenty of animals made it through: frogs, salamanders, lizards and snakes, turtles and crocodiles, mammals, and yes, some dinosaurs—in the guise of birds. Not to mention so many shelled invertebrates and fishes in the oceans, although that could be the subject of another book entirely. So what was it about *T. rex*, *Triceratops*, the sauropods, and their kin that made them a target?

This is a key question. We want to answer it particularly because it's relevant to our modern world. When there is sudden global environmental and climate change, what lives and what dies? It's case studies in the history of life—recorded by fossils, like the end-Cretaceous extinction—that provide critical insight.

The first thing we have to realize is that, although some species did survive the immediate hellfire of the impact and the longer-term climate upheaval, most did not. It's estimated that some 70 percent of species went extinct. That includes a whole lot of amphibians and reptiles and probably the majority of mammals and birds, so it's not simply "dinosaurs died, mammals and birds survived," the line often parroted in textbooks and television documentaries. If not for a few good genes or a few strokes of good luck, our mammalian ancestors might have gone the way of the dinosaurs, and I wouldn't be here typing this book.

There are some things, however, that do seem to distinguish the victims from the survivors. The mammals that lived on were generally smaller than the ones that perished, and they had more omnivorous diets. It seems that being able to scurry around, hide in burrows, and eat a whole variety of different foods was advantageous during the madness of the postimpact world. Turtles and crocodiles fared pretty well compared to other vertebrates, and that is probably because they were able to hide out underwater during those first few hours of bedlam, shielding themselves from the deluge of rock bullets and the earthquakes. Not only that, but their aquatic ecosystems were based on detritus. The critters at the base of their food chain ate decaying plants and other organic matter, not trees, shrubs, and flowers, so their food webs would not have collapsed when photosynthesis was shut down and plants started to die. In fact, plant decay would have just given them much more food.

Dinosaurs had none of these advantages. Most of them were big, and they couldn't easily scamper into burrows to wait out the firestorm. They couldn't hide underwater, either. They were parts of food chains with big plant-eating species at the base, so when the sun was blocked and photosynthesis shut down and plants started to die, they felt the domino effects. Plus, most dinosaurs had fairly specialized diets—they ate meat or particular types of plants, without the flexibility that came with the more adventurous palates of the surviving mammals. And they had other handicaps as well. Many of them were probably warm-blooded or at least had a high metabolism, so they required a lot of food. They couldn't hunker down for months without a meal, like some amphibians and reptiles. They laid eggs, which took between three and six months to hatch, about double the time for birds' eggs. Then, after the eggs hatched, it took dinosaur youngsters many years to grow into adults, a long and tortured adolescence that would have made them particularly vulnerable to environmental changes.

After the asteroid hit, there was probably no one thing that sealed the dinosaurs' fate. They just had a lot of liabilities working against them. Being small, or having an omnivorous diet, or reproducing quickly—none of these things guaranteed survival, but each one increased the odds in what was probably a maelstrom of chance as the Earth devolved into a fickle casino. If life in that moment boiled down to a game of cards, dinosaurs were left holding a dead man's hand.

Some species, however, cashed in on a royal flush. Among them were our mouse-size ancestors, which made it to the other side and soon had the opportunity to build their own dynasty. Then there were the birds. Lots of birds and their close feathery dinosaur cousins died—all of the four-winged and batlike dinosaurs, all of the primitive birds with long tails and teeth. But modern-style birds endured. We don't know why exactly. Maybe it was because their big wings and powerful chest muscles allowed them to literally fly away from the chaos and find safe shelter. Perhaps it was because their eggs hatched quickly, and once out of the nest, the fledglings grew rapidly into adults. It could be that they were specialized for eating seeds—little nuggets of nutrition that can survive in the soil for years, decades, even centuries. Most likely, it was a combination of these assets and others that we have yet to recognize. That and a whole lot of good luck.

After all, so much about evolution—about life—comes down to fate. The dinosaurs got their very chance to rise up after those terrible volcanoes 250 million years ago wiped out nearly every species on Earth, and then they had the good fortune to sail through that second extinction at the end of the Triassic, which felled their crocodile competitors. Now the tables had turned. *T. rex* and *Triceratops* were gone. The sauropods would thunder across the land no more. But let's not forget about those birds—they are dinosaurs, they survived, they are still with us.

The dinosaur empire may be over, but the dinosaurs remain.

EPILOGUE:
AFTER *the* DINOSAURS

EVERY MAY I HEAD OUT to the desert of northwestern New Mexico, not too far from the Four Corners. It's a bit of a break, coming on the heels of exams and paper grading and the usual end-of-semester mania. I usually stay for a couple of weeks, and by the end of my trip, the calm of the empty desert and the spicy food we make every night in camp have succeeded in easing my stress.

It's not a vacation, however. As is usual when I travel these days, I'm here on business—to do what I've spent the last decade doing all over the world, in Polish quarries and on the frigid tidal platforms of Scotland, in the shadows of Transylvanian castles, the outback of Brazil, or the radiating sauna of Hell Creek.

I'm here to find fossils.

Many of these fossils, of course, are dinosaurs. In fact, they're among the last surviving dinosaurs of all, ones that were living about a thousand miles south of Hell Creek during those final few million years of the Cretaceous. They were flourishing at a time when history seemed to be standing still, when it appeared that dinosaurs would keep on ruling the world forever, as they had done for over 150 million years. We find the bones of tyrannosaurs and huge sauropods, the skull domes that pachycephalosaurs used to head-butt each other, the jaws that horned and duck-billed dinosaurs sliced up plants with, and lots of teeth of raptors and other small theropods that skittered around underneath the big boys. So many species, living together in harmony, not a hint that things would soon go horribly wrong.

Truth be told, though, I'm not here for the dinosaurs. That may seem like sacrilege, as I've spent most of my young career on the trail of *T. rex* and *Triceratops*. No, I'm trying to understand what happened after the dinosaurs disappeared—how the Earth healed, a fresh start was made, and a new world was forged.

Most of the candy-striped badlands in this part of New Mexico—in the vast and mostly uninhabited areas of Navajo country, around the towns of Cuba and Farmington—are carved from rocks laid down in rivers and lakes during the first few million years after the asteroid hit. Gone are the tyrannosaur teeth and big chunks of sauropod bone that are so common in the latest Cretaceous rocks of the area, deposited only a few feet below the rocks we're looking at now, which date from the subsequent Paleogene Period (66 to 56 million years ago). There was a sudden change here; the asteroid blew away one world and ushered in another. Many dinosaurs, then all of a sudden, none at all. It's a pattern eerily similar to what Walter Alvarez saw in the forams of the Gubbio gorge.

I walk these dry New Mexican hills with one of my best friends in science: Tom Williamson, a curator at the natural history museum in Albuquerque. Tom has been collecting out here for twenty-five years, starting as a graduate student. He often brings along his twin sons, Ryan and Taylor, who through countless camping trips with their dad have developed a knack for finding fossils that rivals that of almost any paleontologist that I know—even Grzegorz Niedźwiedzki in Poland and Mátyás Vremir in Romania. Other times, Tom comes out here with his students, young Navajos from the surrounding reservations, whose families have lived on this sacred land for generations. And once a year in May, Tom meets me and my students from Edinburgh. Ryan and Taylor—who are now in college—usually tag along, and we have endless fun finding fossils by day and sitting around the campfire at night, telling the sort of

Fieldwork in the badlands of the San Juan Basin of New Mexico, USA.
Tom Williamson.

Me collecting fossils of the mammals that took over from the dinosaurs.

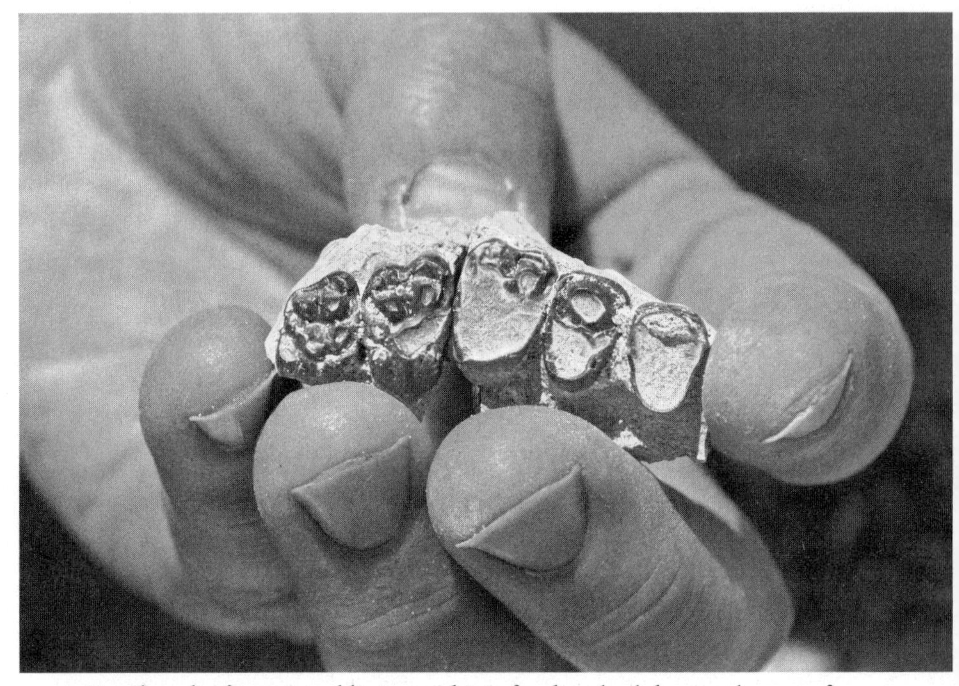

Fossil teeth of a mammal living within a few hundred thousand years of the end-Cretaceous asteroid impact, from New Mexico.

inside jokes that develop after many years of being in the field together.

Tom is blessed with a skill I lack, one that is very useful for a paleontologist. He has a photographic memory. He claims that he doesn't, but that's either false modesty or delusion. Tom can recognize each little hillock and crag in the desert, which all look the same to me. He can recall with precise detail almost every fossil he's ever collected at these sites, which is stunning, because he has collected thousands, perhaps tens of thousands by now.

Fossils are littered all over this landscape, constantly eroding out from the Paleogene rocks. Aside from a few bird bones here and there, these are not dinosaur fossils. They are the jaws and teeth and skeletons of the things that took over from the dinosaurs, the species that went on to start the next great dynasty of Earth history, the dynasty that includes many of the most familiar animals of the modern world, including us.

Mammals.

As you recall, mammals got their start alongside the dinosaurs, born into the violent unpredictability of Pangea over 200 million years ago in the Triassic. But mammals and dinosaurs then went their separate ways. While dinosaurs bested their early crocodile competitors, sailed through the end-Triassic extinction, and then grew to colossal sizes and spread throughout the land, mammals remained in the shadows. They became adept at surviving in anonymity, learning how to eat different foods, hide in burrows, and move around undetected, some even figuring out how to glide through the canopy and others how to swim. All the while, they remained small. No mammal living with dinosaurs got bigger than a badger. They were bit players in the Mesozoic drama.

In New Mexico, however, it is a different story. Those thousands of fossils that Tom can meticulously catalog in his mind belong to a staggering diversity of species. Some are tiny shrew-size insectivores, not too different from the vermin that scurried under the feet of dinosaurs. Others are badger-size burrowers, saber-toothed flesh-eaters, and even cow-size plant-guzzlers. They all lived during the early part of the Paleogene, within half a million years of the asteroid impact.

Already, a mere five hundred thousand years after the most destructive day in the history of Earth, ecosystems had recovered. The temperature was neither nuclear-winter cold nor greenhouse hot. Forests of conifers, gingkos, and an ever-increasing diversity of flowering plants once again towered into the sky. Primitive cousins of ducks and loons loitered near the lakeside, while turtles paddled offshore, oblivious to the crocodiles lurking underneath. But the tyrannosaurs and sauropods and duckbills were no more, replaced by the sudden bounty of mammals that exploded in diversity when presented with an opportunity they had been craving for hundreds of millions of years: a wide open playing field, free of dinosaurs.

Among the mammals that Tom and his crews have discovered is a skeleton of a puppy-size creature called *Torrejonia*. It had gangly limbs and long fingers and toes and probably would have looked, dare I say it, pretty cute and cuddly. It lived about 3 million years after the asteroid hit, but its graceful skeleton doesn't seem that out of place in the world we know today. You can almost envision it leaping through the trees, its skinny toes gripping the branches.

Torrejonia is one of the oldest primates, a fairly close cousin of ours. It is a stark reminder that we—you, me, all of us humans—had ancestors that were there on that terrible day, that saw the rock fall from the sky, that endured the heat and earthquakes and nuclear winter, that eked out passage across the Cretaceous-Paleogene boundary, and then once on the other side, evolved into tree-leapers like *Torrejonia*. Another 60 million years or so of evolution would eventually turn these humble proto-primates into bipedal-walking, philosophizing, book-writing (or reading), fossil-collecting apes. If the asteroid had never hit, if it had never ignited that chain reaction of extinction and evolution, the dinosaurs would probably still be here, and we would not.

There is an even starker reminder, a greater lesson in the dinosaur extinction. What happened at the end of the Cretaceous tells us that even the most dominant animals can go extinct—and quite suddenly. Dinosaurs had been around for over 150 million years when their time of reckoning came. They had endured hardships, evolved superpowers like fast metabolisms and enormous size, and vanquished their rivals so that they ruled an entire planet. Some invented wings so they could fly beyond the bounds of the land; others literally shook the Earth as they walked. There were probably many billions of dinosaurs spread all over the world, from the valleys of Hell Creek to the islands of Europe, that woke up on that day 66 million years ago confident of their undisputed place at the pinnacle of nature.

Then, literally in a split second, it ended.

We humans now wear the crown that once belonged to the dinosaurs. We are confident of our place in nature, even as our actions are rapidly changing the planet around us. It leaves me uneasy, and one thought lingers in my mind as I walk through the harsh New Mexican desert, seeing the bones of dinosaurs give way so suddenly to fossils of *Torrejonia* and other mammals.

If it could happen to the dinosaurs, could it also happen to us?

ACKNOWLEDGMENTS

MY CONTRIBUTION TO THE FIELD of dinosaur research has been relatively recent and relatively small. Like all scientists, I stand on the shoulders of those who came before me, and I'm helped up by those who work alongside me. I hope this book conveys how exciting things are right now in the field of paleontology and how everything we've learned about dinosaurs over the last few decades stems from a communal effort, the work of a diverse group of wonderful people all over the world, men and women, from field volunteers and amateurs to students and professors. There is no way I can thank everyone by name, and I would no doubt forget a bunch of important people if I tried. To all whose names and stories appear in these pages and to everyone I've worked with, thank you for accepting me into the global community of paleontologists and for making the last fifteen years of my life such an incredible ride.

That said, some people deserve special mention here. I've been incredibly privileged to have had three excellent advisors—my undergraduate mentor Paul Sereno at the University of Chicago, Mike Benton when I did my master's at the University of Bristol, and Mark Norell for my PhD at the American Museum of Natural History and Columbia University. I realize now how lucky I was and also how annoying I must have been as a student. These three guys gave me incredible fossils to work on, took me along on fieldwork and research trips around the world, and most important, told me when I was being too ridiculous. I can't help but think that no other young dinosaur researcher has been so fortunate in the mentor department.

I've worked with a lot of people, and most have been very good colleagues—dinosaur paleontologists, at least the modern generation, generally are an agreeable bunch and get along with one another. But some have crossed that line from collaborator to friend, and I would like to particularly thank Thomas Carr and Tom Williamson first and foremost, as well as Roger Benson, Richard Butler, Roberto Candeiro, Tom Challands, Zoltán Csiki-Sava, Graeme Lloyd, Junchang Lü, Octávio Mateus, Sterling Nesbitt, Grzegorz Niedźwiedzki, Dugie Ross, Mátyás Vremir, Steve Wang, and Scott Williams.

I've had a lot of lucky breaks in my young career, none bigger than somehow convincing the University of Edinburgh to hire me while I was finishing up my PhD. Rachel Wood has been the best mentor that a junior faculty member could ever hope for, and she still never lets me pay for coffee, food, beer, or whiskey. Sandy Tudhope, Simon Kelley, Kathy Whaler, Andrew Curtis, Bryne Ngwenya, Lesley Yellowlees, Dave Robertson, Tim O'Shea, and Peter Mathieson have been the best types of bosses—always supportive but never overbearing. Geoff Bromiley, Dan Goldberg, Shasta Marrero, Kate Saunders, Alex Thomas, and the other young guns have made working in Edinburgh fun. Nick Fraser and Stig Walsh have welcomed me into their group at the National Museum of Scotland, and Neil Clark and Jeff Liston have welcomed me into the bigger community of Scottish paleontologists. One of the perks of being on a faculty is that I can advise my own students, and a wonderfully diverse and gifted group has already passed through my lab: Sarah

Shelley, Davide Foffa, Elsa Panciroli, Michela Johnson, Amy Muir, Joe Cameron, Paige dePolo, Moji Ogunkanmi. You probably don't realize how much I have learned from each of you.

Science is hard enough; writing is harder. My two editors, Peter Hubbard at William Morrow in the United States and Robin Harvie in the United Kingdom, have helped mold my anecdotes and rambling into a narrative. A few years ago, Jane von Mehren heard me on the radio and thought that I might have a story to tell; she convinced me to put together a book proposal and has been an amazing agent since then. Also, a big thanks to Esmond Harmsworth and Chelsey Heller at Aevitas, for all of your help negotiating contracts and payments and foreign rights and other fun things. Huge kudos to my buddy, the incomparable artist Todd Marshall, for the original illustrations that enliven my prose, and to my dear friend Mick Ellison, the world's best dino photographer, for letting me use some of his stunning photos. And thanks to my two family lawyers, my dad, Jim, and brother Mike, for making sure each contract was perfect beyond a reasonable doubt.

I've always loved to write, and I've had a lot of people help me along the way. Lonny Cain, Mike Murphy, and Dave Wischnowsky gave me the opportunity to work in the newsroom of my hometown newspaper—the *Times* in Ottawa, Illinois—for four years. The panic of the deadline and the thrill of chasing sources made me learn fast. Many folks published my (often awful) teenage writings on dinosaurs in their magazines and on their websites, especially Fred Bervoets, Lynne Clos, Allen Debus, and Mike Fredericks. More recently, Kate Wong at *Scientific American*, Richard Green at Quercus, Florian Maderspacher at *Current Biology*, and Stephen Khan, Steven

Vass, and Akshat Rathi at *The Conversation* have given me both a platform and tough editorial love. When I got started writing this book, Neil Shubin (one of my undergrad profs) and Ed Yong both gave very helpful advice.

I'd like to thank many funding agencies—too numerous to name here—for regularly turning down my grant applications, giving me ample time and freedom to write this book. On the other hand, my sincere gratitude goes to the National Science Foundation and the Bureau of Land Management (and the US taxpayers who fund them), the National Geographic Society, the Royal Society and the Leverhulme Trust in the UK, and the EU-funded European Research Council and Marie Skłodowska-Curie Actions (and the European governments and taxpayers who fund them) for supporting me. I've also received many small grants from various sources and ample support from the American Museum of Natural History and the University of Edinburgh.

I have the best family of anyone I know. My parents, Jim and Roxanne, let me drag them to museums on family vacations and made sure that I could study paleontology in college. My brothers, Mike and Chris, went along with it. Nowadays, my wife, Anne, goes along with it too. She tolerates my absences on fieldwork, the times I need to sneak upstairs to write, and the various dinosaur-obsessed houseguests and pub mates that I inevitably attract. She even read through drafts of this book, even though she has no interest in dinosaurs whatsoever. Much luv! Anne's parents, Peter and Mary, have let me spend a lot of time in their house in Bristol, England, a quiet place to write. I have some other cool in-laws too: my wife's sister Sarah and Mike's wife, Stephenie.

Finally, my thanks to all of the unsung heroes, the folks who usually remain anonymous but without whom our field would grind to extinction. The fossil preparators, the field technicians, the undergraduate assistants, the university secretaries and administrators, the patrons who visit museums and donate to universities, the science journalists and feature writers, the artists and photographers, the journal editors and peer reviewers, the amateur collectors who do good and donate their fossils to museums, the folks who administer public lands and process our permits (particularly my friends in the Bureau of Land Management, the Scottish Natural Heritage, and the Scottish government), the politicians and federal agencies who support science (and stand up to those who do not), the taxpayers and voters who support research, all the science teachers at all levels, and so many more.

NOTES
ON SOURCES

My main source of information for this book was personal experience—the fossils that I've studied, the fieldwork I've done, the museum collections I've visited, and many discussions with scientific colleagues and friends. In writing this book, I picked over many of the scientific papers I've written for various journals, my textbook *Dinosaur Paleobiology* (Hoboken, NJ: Wiley-Blackwell, 2012), and pop-science pieces I've written for *Scientific American* and the *Conversation*. The following notes mention some supplementary material and sources that I used, which I direct you to for more information.

PROLOGUE: THE GOLDEN AGE OF DISCOVERY

I tell the story of my journey to Jinzhou to study *Zhenyuanlong* in one of my pieces for *Scientific American*, "Taking Wing," vol. 316, no. 1 (Jan. 2017): 48–55. Junchang Lü and I described *Zhenyuanlong* in a 2015 paper in *Scientific Reports* 5, article no. 11775.

CHAPTER 1: THE DAWN OF THE DINOSAURS

There are two well-written pop-science books on the end-Permian extinction, one by my former master's advisor Mike Benton (*When Life Nearly Died: The Greatest Mass Extinction of*

All Time, Thames & Hudson, 2003) and the other by the great Smithsonian paleontologist Douglas Erwin (*Extinction: How Life on Earth Nearly Ended 250 Million Years Ago*, Princeton University Press, 2006). Zhong-Qiang Chen and Mike Benton wrote a short semitechnical review of the extinction and subsequent recovery for *Nature Geoscience* (2012, 5: 375–83). Updated information on the timing and nature of the volcanic eruptions that caused the extinction was published by Seth Burgess and colleagues: *Proceedings of the National Academy of Sciences USA* 111, no. 9 (Sept. 2014): 3316–21; and *Science Advances* 1, no. 7 (Aug. 2015): e1500470. Some excellent technical papers on the extinction have been written by Jonathan Payne, Peter Ward, Daniel Lehrmann, Paul Wignall, and my Edinburgh colleague Rachel Wood and her PhD student Matt Clarkson, whom I once roped into filling in on a faculty committee only a few days after he finished his thesis.

Grzegorz Niedźwiedzki has published numerous papers on the Permian-Triassic tracks of the Holy Cross Mountains of Poland. In many of these, he is joined by his friends Tadeusz Ptaszyński, Gerard Gierliński, and Grzegorz Pieńkowski of the Polish Geological Institute. Pieńkowski is a charming chap who was active in the Solidarity movement in the 1980s and was rewarded for his political activism with a consul general post in Australia when democrats assumed power after the fall of Communism. He kindly opened his guesthouse to us and plied us with kielbasa when we were traveling through the northeastern Polish lake district on our way to try to find fossils in Lithuania. Our joint work on *Prorotodacylus* and early dinosauromorph tracks was first published in 2010 as Stephen L. Brusatte, Grzegorz Niedźwiedzki, and Richard J. Butler, "Footprints Pull Origin and Diversification of Dinosaur Stem Lineage Deep into Early Triassic," *Proceedings of the Royal Society of London Series B*, 278 (2011): 1107–13, and later as a longer monograph with Grzegorz as the lead author in *Anatomy, Phylogeny, and Palaeobiology of Early Archosaurs and Their Kin*, ed. Sterling J. Nesbitt, Julia B. Desojo, and Randall B. Irmis (Geological Society of London Special Publications no. 379, 2013), pp. 319–51. Important work on Triassic tracks from other parts of the world has been published by Paul Olsen, Hartmut Haubold, Claudia Marsicano, Hendrik Klein, Georges Gand, and Georges Demathieu.

The family tree of dinosaurs and close relatives that I developed during my master's degree work was published as "The Higher-Level Phylogeny of Archosauria," *Journal of Systematic Palaeontology* 8, no. 1 (Mar. 2010): 3–47.

The chapter focuses on the tracks of early dinosauromorphs that I've studied and only briefly mentions the skeletal fossils of these animals. There is a growing record of skeletons belonging to species like *Silesaurus* (the "intriguing new reptile fossils" found in Silesia alluded to in the text, studied by Jerzy Dzik, the "very senior Polish professor"), *Lagerpeton*, *Marasuchus*, *Dromomeron*, and *Asilisaurus*. A semitechnical review of these animals was published by Max Langer and colleagues in *Anatomy, Phylogeny, and Palaeobiology of Early Archosaurs and Their Kin*, pp. 157–86. *Nyasasaurus*, the puzzling creature that may be the oldest dinosaur or merely a close cousin, was described by Sterling Nesbitt and colleagues in *Biology Letters* 9 (2012), no. 20120949.

Cherry Lewis's biography of Arthur Holmes (*The Dating Game: One Man's Search for the Age of the Earth*, Cambridge University Press, 2000) is a good introduction to the concept of radiometric dating, the history of its discovery, and how it is used to date rocks. The sticky subject of dating Triassic rocks is discussed in an important paper by Claudia Marsicano, Randy Irmis, and colleagues (*Proceedings of the National Academy of Sciences USA*, 2015, doi: 10.1073/pnas.1512541112).

Paul Sereno, Alfred Romer, José Bonaparte, Osvaldo Reig, Oscar Alcober, and their students and colleagues have written many papers on the Ischigualasto dinosaurs and the animals living alongside them. The single best source of information is the 2012 Memoir of the Society of Vertebrate Paleontology, *Basal Sauropodomorphs and the Vertebrate Fossil Record of the Ischigualasto Formation (Late Triassic: Carnian–Norian) of Argentina*, which includes a historical review of the Ischigualasto expeditions and a detailed anatomical description of *Eoraptor*, both written by Sereno.

Two interesting developments were published right as this book was going to press. First, the Ischigualasto plant-eating *Pisanosaurus*, which I discuss as an early member of the ornithischian dinosaur lineage, was redescribed and reclassified as a non-dinosaur dinosauromorph closely related to *Silesaurus* (F. L. Agnolin and S. Rozadilla, *Journal of Systematic Palaeontology*, 2017, http://dx.doi.org/10.1080/14772019.2017.1352623). So it is possible that there are currently no good ornithischian fossils from the entire Triassic Period. Secondly, Cambridge PhD student Matthew Baron and colleagues published a new family tree of dinosaurs, placing theropods and ornithischians in their own group (Ornithoscelida) exclusive of sauropods (*Nature*, 2017, 543: 501–6). This is an exciting but controversial idea. I was part of a team, led by Max Langer, that reassessed Baron et al.'s data set and argued for the more traditional ornithischian-saurischian subdivision of dinosaurs (*Nature*, 2017, 551: E1–E3, doi:10.1038/nature24011). This will surely be a huge subject of debate for many years.

CHAPTER 2: DINOSAURS RISE UP

There are several reviews concerning the rise of dinosaurs during the Triassic. I wrote one with several of my colleagues, including Sterling Nesbitt and Randy Irmis of the Rat Pack: Brusatte et al., "The Origin and Early Radiation of Dinosaurs," *Earth-Science Reviews* 101, no. 1–2 (July 2010): 68–100. Others have been written by Max Langer and various colleagues: Langer et al., *Biological Reviews* 85 (2010): 55–110; Michael J. Benton et al., *Current Biology* 24, no. 2 (Jan. 2014): R87–R95; Langer, *Palaeontology* 57, no. 3 (May 2014): 469–78; Irmis, *Earth and Environmental Science Transactions of the Royal Society of Edinburgh*, 101, no. 3–4 (Sept. 2010): 397–426; and Kevin Padian, *Earth and Environmental Science Transactions of the Royal Society of Edinburgh* 103, no. 3–4 (Sept. 2012): 423–42.

Two excellent semitechnical books on the Triassic Period and how dinosaurs fit into the larger "assembly of modern ecosystems" have been written by my friend down the road at the National Museum of Scotland, Nick Fraser. In 2006 Nick published *Dawn of the Dinosaurs: Life in the Triassic* (Indiana University Press), and in 2010 he joined Hans-Dieter Sues in writing *Triassic Life on Land: The Great Transition* (Columbia University Press). These books are both richly illustrated (the former by the great paleoartist Doug Henderson) and contain references to most of the important primary literature on Triassic vertebrate evolution. The best maps of ancient Pangea—studiously drawn based on many lines of geological evidence that can trace ancient shorelines and determine the positions of land millions of years ago—have been produced by Ron Blakey and Christopher Scotese. Throughout the book, I have relied on these extensively when explaining the breakup of Pangea.

We have published a few papers on our excavations in Portugal, including a detailed description of the *Metoposaurus* skeletons found in the mass grave: Brusatte et al., *Journal of Vertebrate Paleontology* 35, no. 3, article no. e912988 (2015): 1–23; and a description of the phytosaur that lived with the "Super Salamanders": Octávio Mateus et al., *Journal of Vertebrate Paleontology* 34, no. 4 (2014): 970–75. The German geology student who found the first Triassic specimens in the Algarve was Thomas Schröter, and the "obscure" paper that described the fossils he found was written by Florian Witzmann and Thomas Gassner, *Alcheringa* 32, no. 1 (Mar. 2008): 37–51.

The Rat Pack—Randy Irmis, Sterling Nesbitt, Nate Smith, Alan Turner, and their colleagues—have published numerous papers on the specimens they found at Ghost Ranch, the paleoenvironment of the area, and how their finds fit into the global context of Triassic dinosaur evolution. Among the most important are Nesbitt, Irmis, and William G. Parker, *Journal of Systematic Palaeontology* 5, no. 2 (May 2007): 209–43; Irmis et al., *Science* 317, no. 5836 (July 20, 2007): 358–61; and Jessica H. Whiteside et al., *Proceedings of the National Academy of Sciences USA* 112, no. 26 (June 30, 2015): 7909–13. Edwin Colbert comprehensively described the Ghost Ranch *Coelophysis* skeletons in his 1989 monograph *The Triassic Dinosaur Coelophysis*, *Museum of Northern Arizona Bulletin* 57: 1–160, and he recounted the story of their discovery in many of his gripping popular books on dinosaurs. Martín Ezcurra's paper on *Eucoelophysis* was published in *Geodiversitas* 28, no. 4: 649–84. Sterling Nesbitt described *Effigia* in a short paper in 2006 in *Proceedings of the Royal Society of London, Series B*, vol. 273 (2006): 1045–48, and later as a monograph in the *Bulletin of the American Museum of Natural History* 302 (2007): 1–84.

My work on the morphological disparity of Triassic dinosaurs and pseudosuchians was published in two papers in 2008: Brusatte et al., "Superiority, Competition, and Opportunism in the Evolutionary Radiation of Dinosaurs," *Science* 321, no. 5895 (Sept. 12, 2008): 1485–88; and Brusatte et al., "The First 50 Myr of Dinosaur Evolution," *Biology Letters* 4: 733–36. These papers were cowritten with Mike Benton, Marcello Ruta, and Graeme Lloyd, my MSc supervisors at the University of Bristol and some of my most trusted colleagues in the field today. The publications that inspired me, written by Bakker and Charig, are cited and discussed in those papers. Many invertebrate paleontologists helped develop the standard morphological disparity methods, especially Mike Foote (who was on the faculty of my undergraduate institution, the University of Chicago, but whom I was unfortunately never able to take a course with) and Matt Wills, and I have also cited their papers extensively in my work.

The name Mike Benton pops up a lot in this section. I've said less about Mike in the main text than my other two academic advisors, Paul Sereno and Mark Norell, probably because I was in Bristol for too short of time to accumulate the sort of juicy stories that fit into the way I've chosen to write this narrative. But that is no reflection of Mike. He is a scientific superstar whose studies of vertebrate evolution and popular textbooks (particularly *Vertebrate Palaeontology*, which has gone through several editions with Wiley-Blackwell, most recently in 2014) have set the pulse for the entire field of vertebrate paleontology for decades. But despite the wide regard in which he is held, he is a humble man who is well loved for being a helpful supervisor to dozens of graduate students.

CHAPTER 3: DINOSAURS BECOME DOMINANT

The books *Dawn of the Dinosaurs: Life in the Triassic* and *Triassic Life on Land: The Great Transition*, both cited above, in the notes to chapter 2, provide excellent overviews of the end-Triassic extinction. Some of the topics of this chapter are also discussed in the review papers on early dinosaur evolution used as sources for chapter 2.

The lava erupting at the end of the Triassic created a huge amount of basaltic rock (including the Palisades of New Jersey), which covers part of four continents today. This is referred to as the Central Atlantic Magmatic Province (or CAMP), and has been well described by Marzoli and colleagues in *Science* 284, no. 5414 (Apr. 23, 1999): 616–18. The timing of the CAMP eruptions has been studied by Blackburn and colleagues, including Paul Olsen, in *Science* 340, no. 6135 (May 24, 2013): 941–45, and it is their work that shows that the eruptions took place in four large pulses over six hundred thousand years. Work by Jessica Whiteside, our friend from Portugal and Ghost Ranch, has shown that the extinctions on land and in the sea happened at the same time at the end of the Triassic, and that the first hints of extinction are synchronous with the first lava flows in Morocco. See *Proceedings of the National Academy of Sciences USA* 107, no. 15 (Apr. 13, 2010): 6721–25. Paul Olsen was also part of this research, as he was Whiteside's PhD supervisor at Columbia University.

Changes across the Triassic-Jurassic boundary in atmospheric carbon dioxide, global temperature, and plant communities have been studied by, among others, Jennifer McElwain and colleagues in *Science* 285, no. 5432 (Aug. 27, 1999): 1386–90, and *Paleobiology* 33, no. 4 (Dec. 2007): 547–73; Claire M. Belcher et al., *Nature Geoscience* 3 (2010): 426–29; Margret Steinthorsdottir et al., *Palaeogeography, Palaeoclimatology, Palaeoecology* 308 (2011): 418–32; Micha Ruhl and colleagues, *Science* 333, no. 6041 (July 22, 2011): 430–34; and Nina R. Bonis and Wolfram M. Kürschner, *Paleobiology* 38, no. 2 (Mar. 2012): 240–64.

Paul Olsen has been publishing on the rift basins and fossils of eastern North America since just a few years after his teenage hijinks. He has written two technical overviews of the Pangean rift basin system (which geologists call the Newark Supergroup), both with Peter LeTourneau, *The Great Rift Valleys of Pangea in Eastern North America*, vols. 1–2 (Columbia University Press, 2003), and a very useful review paper on the subject in the *Annual Review of Earth and Planetary Sciences* 25 (May 1997): 337–401. In 2002 Olsen published an important paper summarizing his years of work on footprints, which presented evidence for the rapid radiation of dinosaurs after the end-Triassic extinction: *Science* 296, no. 5571 (May 17, 2002): 1305–7.

There is a huge literature on sauropod dinosaurs. One of the best technical books on these iconic dinosaurs was edited by Kristina Curry Rogers and Jeff Wilson: *The Sauropods: Evolution and Paleobiology* (University of California Press, 2005). A good technical summary was written by Paul Upchurch, Paul Barrett, and Peter Dodson for the second edition of the classic scholarly dinosaur encyclopedia, *The Dinosauria* (University of California Press, 2004), and I wrote a less technical review of the group in my 2012 textbook, *Dinosaur Paleobiology* (Hoboken, NJ: Wiley-Blackwell). My early-career colleagues Phil Mannion and Mike D'Emic have recently been doing a lot of excellent descriptive work on sauropods, along with their advisors Upchurch, Barrett, and Wilson.

We described our sauropod dinosaur trackways from Skye in 2016: Brusatte et al., *Scottish Journal of Geology* 52: 1–9. Some of the earlier fragmentary records of Scottish sauropods were presented by my Glasgow buddy Neil Clark and Dugie Ross in the *Scottish Journal of Geology* 31 (1995): 171–76; by my incomparable Scottish nationalist comrade Jeff Liston, *Scottish Journal of Geology* 40, no. 2 (2004): 119–22; and by Paul Barrett, *Earth and Environmental Science Transactions of the Royal Society of Edinburgh* 97: 25–29.

Calculating the body weights of dinosaurs has been the focus of numerous studies. A pioneering work by J. F. Anderson and colleagues was the first to recognize the relationship between long-bone thickness (technically, circumference) and body weight (technically, mass) in modern and extinct animals: *Journal of Zoology* 207, no. 1 (Sept. 1985): 53–61. More recent work by Nic Campione, David Evans, and colleagues has refined this approach: *BMC Biology* 10 (2012): 60; and *Methods in Ecology and Evolution* 5 (2014): 913–23. These methods have been used to estimate the masses of nearly all dinosaurs by Roger Benson and coauthors: *PLoS Biology* 12, no. 5 (May 2014): e1001853.

The photogrammetry-based method for estimating mass was pioneered by Karl Bates and his PhD supervisors Bill Sellers and Phil Manning in *PLoS ONE* 4, no. 2 (Feb. 2009): e4532, and has since been expanded on in several publications, including Sellers et al., *Biology Letters* 8 (2012): 842–45; Brassey et al., *Biology Letters* 11 (2014): 20140984; and Bates et al., *Biology Letters* 11 (2015): 20150215). Peter Falkingham has published a primer on how to collect photogrammetric data in *Palaeontologica Electronica* 15 (2012): 15.1.1T. The work on sauropods that I took part in, led by Karl, Peter, and Viv Allen, was published in *Royal Society Open Science* 3 (2016): 150636.

It is worth noting that both of these methods—equations based on long-bone circumference and photogrammetric models—do have sources of error. These errors become larger for larger dinosaurs, particularly as the methods cannot be validated in modern animals that are anywhere near the size of sauropods. The original publications cited above extensively discuss these sources of error and, in many cases, present a range of plausible body masses for each dinosaur species based on this understanding of uncertainty.

The biology and evolution of sauropods are subjects of a fascinating collection of research papers gathered together into the book *Biology of the Sauropod Dinosaurs: Understanding the Life of Giants,* ed. Nicole Klein and Kristian Remes (Indiana University Press, 2011). A chapter in this book, written by Oliver Rauhut and colleagues, discusses in detail the evolution of the sauropod body plan: how all the characteristic features of the group came together over millions of years. The question of why sauropods were able to get so big was recently tackled in an excellent, accessible review paper on sauropod biology written by Martin Sander and a team of researchers who studied this mystery over many years, funded by a large German research grant: *Biological Reviews* 86 (2011): 117–55.

CHAPTER 4: DINOSAURS AND DRIFTING CONTINENTS

For information on Zallinger's mural, check out Richard Conniff's *House of Lost Worlds: Dinosaurs, Dynasties, and the Story of Life on Earth* (Yale University Press, 2016) or Rosemary Volpe's *The Age of Reptiles: The Art and Science of Rudolph Zallinger's Great Dinosaur Mural at Yale* (Yale Peabody Museum, 2010). Better yet, go see the mural for yourself at the Peabody Museum if you have the chance. It's a stunning work of art.

There are many popular accounts of the Cope-Marsh Bone Wars, but for a scholarly and matter-of-fact version, I recommend John Foster's excellent book *Jurassic West: The Dinosaurs of the Morrison Formation and Their World* (Indiana University Press, 2007). Foster has spent decades excavating dinosaurs throughout the American West, and his book is a masterful summary of the Morrison dinosaurs, the world they lived in, and the history of their discovery. I relied on this book as my primary source for the historical information in this chapter. The book cites numerous primary sources, including the many research papers that Cope and Marsh published during their hyperactive feud.

The story of Big Al is based on a report by then University of Wyoming and now BLM paleontologist Brent Breithaupt, written for the National Park Service and published as "The Case of 'Big Al' the *Allosaurus*: A Study in Paleodetective Partnerships," in V. L. Santucci and L. McClelland, eds., *Proceedings of the 6th Fossil Resource Conference* (National Park Service, 2001), 95–106.

Interesting studies have been published on the body size (Bates et al., *Palaeontologica Electronica*, 2009, 12: 3.14A) and pathologies (Hanna, *Journal of Vertebrate Paleontology*, 2002, 22: 76–90) of Big Al, and the computer modeling study of *Allosaurus* feeding that I refer to was published by Emily Rayfield and colleagues (*Nature*, 2001, 409: 1033–37). Information about Kirby Siber was gleaned from a profile in *Rocks & Minerals Magazine*, written by John S. White (2015, 90: 56–61). For a balanced take on the subject of commercial fossil collecting and the selling of dinosaur fossils, Heather Pringle's article in *Science* (2014, 343: 364–67) is a good place to start.

There are many great research articles on the sauropods of the Morrison Formation. The best place to start is with the sauropod chapter in the academic textbook *The Dinosauria*, written by sauropod experts Paul Upchurch, Paul Barrett, and Peter Dodson (University of California Press, 2004). Over the past two decades, there has been considerable debate about how different sauropods held their necks, which I summarize in my textbook *Dinosaur Paleobiology*, with citations to the relevant literature, much of which has been written by Kent Stevens and Michael Parrish. There has also been a great deal of work on sauropod feeding habits, with some of the more important papers by Upchurch and Barrett. These are discussed and summarized in both my textbook and the 2011 Sander et al. paper on sauropods cited at the end of the notes for chapter 3 above. More recently, Upchurch, Barrett, Emily Rayfield, and their PhD students David Button and Mark Young have done groundbreaking computer modeling work aimed at understanding how different sauropods fed (Young et al., *Naturwissenschaften*, 2012, 99: 637–43; Button et al., *Proceedings of the Royal Society of London, Series B*, 2014, 281: 20142144).

The chapters in *The Dinosauria* are good sources of information on the Late Jurassic dinosaurs of other continents. The now famous Late Jurassic dinosaurs of Portugal have been studied extensively by Octávio Mateus, my friend and co-excavator of the "SuperSalamander" bone bed whom we met earlier. For a review, see Antunes and Mateus, *Comptes Rendus Palevol* 2 (2003): 77–95. The Late Jurassic dinosaurs of Tanzania were excavated during a series of remarkable German-led expeditions in the early 1900s, which are described in the detailed historical account of Gerhard Maier in his book *African Dinosaurs Unearthed: The Tendaguru Expeditions* (Indiana University Press, 2003).

My primary source for the changes that occurred across the Jurassic-Cretaceous boundary is an excellent review paper by Jonathan Tennant and coauthors (*Biological Reviews*, 2016, 92 (2017): 776-814). I was one of the peer reviewers of this paper, and of the many hundreds of manuscripts I've reviewed, this one may be the one that I learned the most from. Jon did this work as a PhD student in London. Those Internet geeks among you may recognize him as a prolific tweeter and a very passionate communicator of science through blogs and social media.

There have been many profiles of Paul Sereno in books, magazines, and newspapers. Some of those I wrote back in the late 1990s and early 2000s during my fanboy days, but I won't give the specifics here just to make it a little extra difficult for anyone who wants to track down those embarrassing excuses for journalism. One day Paul will probably (I hope!) write his own story, but in the meantime, there is extensive information on his expeditions and discoveries on his lab website (paulsereno.org). Some of his more important African discoveries include the following, with short citations to the relevant scientific papers in parentheses: *Afrovenator* (*Science*, 1994, 266: 267–70); *Carcharodontosaurus saharicus* and *Deltadromeus* (*Science*, 1996, 272: 986–91); *Suchomimus* (*Science*, 1998, 282: 1298–1302); *Jobaria* and *Nigersaurus* (*Science*, 1999, 286: 1342–47); *Sarcosuchus* (*Science*, 2001, 294: 1516–19); *Rugops* (*Proceedings of the Royal Society of London Series B*, 2004, 271: 1325–30). Paul and I described *Carcharodontosaurus iguidensis* together in 2007 (Brusatte and Sereno, *Journal of Vertebrate Paleontology* 27: 902–16) and *Eocarcharia* a year later (Sereno and Brusatte, *Acta Palaeontologica Polonica*, 2008, 53: 15–46).

There is a huge literature of textbooks and how-to guides on the subject of building family trees (phylogenies) using cladistics. The theory behind the methods was developed by the German entomologist Willi Hennig, who outlined his ideas in a paper (*Annual Review of Entomology*, 1965, 10: 97–116) and a landmark book, *Phylogenetic Systematics* (University of Illinois Press, 1966). Those works can be quite dense, but more approachable are textbooks by Ian Kitching et al. (*Cladistics: The Theory and Practice of Parsimony Analysis*, Systematics Association, London, 1998), Joseph Felsenstein (*Inferring Phylogenies*, Sinauer Associates, 2003), and Randall Schuh and Andrew Brower (*Biological Systematics: Principles and Applications*, Cornell University Press, 2009). I also give a general explanation, using dinosaurs as an example, in the phylogeny chapter of my textbook *Dinosaur Paleobiology*.

I published my family tree of carcharodontosaurids (and their allosaur relatives) in 2008, in a paper written with Paul Sereno (*Journal of Systematic Palaeontology* 6: 155–82). I published an updated version the following year, when I joined other colleagues to name and describe the first Asian carcharodontosaurid, *Shaochilong* (Brusatte et al., *Naturwissenschaften*, 2009, 96: 1051–58). One of the coauthors on that paper was Roger Benson, who like me was a student at the time. Roger and I became fast friends, traveled to many museums together (including an incredible trip to China in 2007), and collaborated on several research projects on carcharodontosaurids and other allosaurs, among them a monographic description of the English carcharodontosaur *Neovenator* (Brusatte, Benson, and Hutt, *Monograph of the Palaeontographical Society*, 2008, 162: 1–166). Roger invited me to take part in a further study of carcharodontosaurid/allosaur/theropod phylogeny, on which he did the vast majority of work (Benson et al., *Naturwissenschaften*, 2010, 97: 71–78).

CHAPTER 5: THE TYRANT DINOSAURS

This chapter is something of an expanded version of an article that I wrote for *Scientific American* of May 2015 (312: 34–41) on the story of tyrannosaur evolution. That article took its inspiration from a review paper on tyrannosaur genealogy and evolution that I published with several colleagues in 2010 (Brusatte et al., *Science*, 329: 1481–85). Both are good general sources of information on tyrannosaurs, as is Thomas Holtz's chapter in the academic textbook *The Dinosauria* (University of California Press, 2004).

Junchang Lü and I described *Qianzhousaurus sinensis* (Pinocchio rex) in a paper in 2014 (Lü et al., *Nature Communications* 5: 3788). The story of its discovery was recounted in a *New York Times* article by Didi Kirsten Tatlow (sinosphere.blogs.nytimes.com/2014/05/08/pinocchio-rex-chinas-new-dinosaur). The "weird tyrannosaur" *Alioramus* that I studied, which led to Junchang asking me to help him study *Qian-*

zhousaurus, was described in a series of papers: Brusatte et al., *Proceedings of the National Academy of Sciences USA* 106 (2009): 17261–66; Bever et al., *PLoS ONE* 6, no. 8 (Aug. 2011): e23393; Brusatte et al., *Bulletin of the American Museum of Natural History* 366 (2012): 1–197; Bever et al., *Bulletin of the American Museum of Natural History* 376 (2013): 1–72; and Gold et al., *American Museum Novitates* 3790 (2013): 1–46.

For nearly a decade, I've been studying the genealogy of tyrannosaurs and building ever-larger family trees as new tyrannosaur fossils are found. This work has been done in partnership with my good friend and colleague Thomas Carr of Carthage College in Kenosha, Wisconsin. We published the first version of the family tree in the 2010 *Science* review paper mentioned above. In 2016 we published a fully revamped version (Brusatte and Carr, *Scientific Reports* 6: 20252). It is the 2016 family tree that provides the framework for the discussion of evolution in this chapter.

The discovery of *T. rex* has been recounted in many popular and scientific accounts. The best source of information on Barnum Brown and his great discovery is a biography of Brown that Lowell Dingus and Mark Norell, my PhD advisor, published in 2011 (*Barnum Brown: The Man Who Discovered Tyrannosaurus rex*, University of California Press). Lowell's quote that I use in the chapter comes from an American Museum of Natural History website devoted to the book. There is an excellent biography of Henry Fairfield Osborn written by Brian Rangel, which I used for information on his life (*Henry Fairfield Osborn: Race and the Search for the Origins of Man*, Ashgate Publishing, Burlington, VT, 2002).

Sasha Averianov described *Kileskus* in a paper in 2010 (Averianov et al., *Proceedings of the Zoological Institute RAS*, 314: 42–57). Xu Xing and colleagues described *Dilong* in 2004 (Xu et al., *Nature* 431: 680–84),; *Guanlong* in 2006 (Xu et al., *Nature* 439: 715–18), and *Yutyrannus* in 2012 (Xu et al., *Nature* 484: 92–95). The description of *Sinotyrannus* was written by Ji Qiang and colleagues (Ji et al., *Geological Bulletin of China*, 2009, 28: 1369–74). Roger Benson and I named *Juratyrant* (Brusatte and Benson, *Acta Palaeontologica Polonica*, 2013, 58: 47–54), based on a specimen that Roger described a few years earlier (Benson, *Journal of Vertebrate Paleontology*, 2008, 28: 732–50). *Eotyrannus*, from the beautiful Isle of Wight in England, was named and described by Steve Hutt and colleagues (Hutt et al., *Cretaceous Research*, 2001, 22: 227–42).

Our paper naming and describing *Timurlengia* from the middle Cretaceous of Uzbekistan was published in 2016 (Brusatte et al., *Proceedings of the National Academy of Sciences USA* 113: 3447–52). Also joining Sasha, Hans, and me were my master's student Amy Muir (who processed the CT scan data) and Ian Butler (fellow University of Edinburgh faculty, who custom-built the CT scanner we used to study the fossil). For information on the carcharodontosaurs that were still holding tyrannosaurs down during the middle Cretaceous, check out the papers describing *Siats* (Zanno and Makovicky, *Nature Communications*, 2013, 4: 2827), *Chilantaisaurus* (Benson and Xu, *Geological Magazine*, 2008, 145: 778–89), *Shaochilong* (Brusatte et al., *Naturwissenschaften*, 2009, 96: 1051–58), and *Aerosteon* (Sereno et al., *PLoS ONE*, 2008, 3, no. 9: e3303).

CHAPTER 6: THE KING OF THE DINOSAURS

The story I open with is conjectural, of course, but the details are based on actual fossil discoveries (described later in the chapter, and referenced below), with a dose of speculation about how *T. rex*, *Triceratops*, and duck-billed dinosaurs would have behaved.

For general background on *T. rex*—its size, body features, habitat, and age—please refer to the general references on tyrannosaurs cited in the previous chapter. Body mass estimates come from the previously cited paper on dinosaur body size evolution by Roger Benson and colleagues.

There is a wealth of literature on the feeding habits of *T. rex*. The information on daily food intake comes from two important papers on the subject: one written by James Farlow (*Ecology*, 1976, 57: 841–57) and the other by Reese Barrick and William Showers (*Palaeontologia Electronica*, 1999, vol. 2, no. 2). The idea that *T. rex* was a scavenger, which frustrates the hell out of many dinosaur paleontologists (especially me) whenever it rears its head to make its latest round in the press, has been thoroughly debunked by one of the most knowledgeable and enthusiastic tyrannosaur experts around, Thomas Holtz, in *Tyrannosaurus rex: The Tyrant King* (Indiana University Press, 2008). The fossil *Edmontosaurus* bones with a *T. rex* tooth embedded inside were described by a team led by Robert DePalma (*Proceedings of the National Academy of Sciences USA*, 2013, 110: 12560–64). The famous bone-filled tyrannosaur dung was described by Karen Chin and colleagues (*Nature*, 1998, 393: 680–82), and the bony stomach contents were described by David Varricchio (*Journal of Paleontology*, 2001, 75: 401–6).

Puncture-pull feeding in tyrannosaurs has been studied in detail by Greg Erickson and his team, who have published several important papers on the subject (e.g., Erickson and Olson,

Journal of Vertebrate Paleontology, 1996, 16: 175–78; Erickson et al., *Nature*, 1996, 382: 706–8). Other important studies have been presented by Mason Meers (*Historical Biology*, 2002, 16: 1–2), François Therrien and colleagues (in *The Carnivorous Dinosaurs*, Indiana University Press, 2005), and Karl Bates and Peter Falkingham (*Biology Letters*, 2012, 8: 660–64). Emily Rayfield's most salient publications on tyrannosaur skull construction and biting behavior were two papers published in the mid-2000s (*Proceedings of the Royal Society of London Series B*, 2004, 271: 1451–59; and *Zoological Journal of the Linnean Society*, 2005, 144: 309–16). She has also written a very helpful primer on finite element analysis (*Annual Review of Earth and Planetary Sciences*, 2007, 35: 541–76).

John Hutchinson and his collaborators have written many research papers on tyrannosaur locomotion. Chief among these are articles in *Nature* (2002, 415: 1018–21), *Paleobiology* (2005, 31: 676–701), *Journal of Theoretical Biology* (2007, 246: 660–80), and *PLoS ONE* (2011, 6, no. 10: e26037). Working with Matthew Carrano, John published an important study on *T. rex* pelvic and hind limb musculature (*Journal of Morphology*, 2002, 253: 207–28). John has also written a general primer on studying locomotion in dinosaurs (in the *Encyclopedia of Life Sciences*, Wiley-Blackwell, 2005), but you can find his best writing on his always-entertaining blog (https://whatsinjohnsfreezer.com/).

The efficient lung of modern birds, and how it works, is described in more detail in my book *Dinosaur Paleobiology*. There are also a few specialist papers on the subject worth examining (e.g., Brown et al., *Environmental Health Perspectives*, 1997, 105: 188–200; and Maina, *Anatomical Record*, 2000, 261: 25–44). The fossilized evidence for air sacs in dinosaur bones—referred to technically as pneumaticity—was expertly studied by Brooks Britt during his PhD work (Britt, 1993, PhD thesis, University of Calgary). More recently, important work on the subject has been presented by Patrick O'Connor and colleagues (*Journal of Morphology*, 2004, 261: 141–61; *Nature*, 2005, 436: 253–56; *Journal of Morphology*, 2006, 267: 1199–1226; *Journal of Experimental Zoology*, 2009, 311A: 629–46), by Roger Benson and collaborators (*Biological Reviews*, 2012, 87: 168–93), and by Mathew Wedel (*Paleobiology*, 2003, 29: 243–55; *Journal of Vertebrate Paleontology*, 2003, 23: 344–57).

Sara Burch's research on tyrannosaur arms was described in her PhD thesis (Stony Brook University, 2013) and has been presented at annual meetings of the Society of Vertebrate Paleontology. It is currently awaiting full publication.

Phil Currie and his crew wrote several papers on the *Albertosaurus* mass grave, which fill a special issue of the *Canadian Journal of Earth Sciences* (2010, vol. 47, no. 9). Phil's work on pack hunting in *Albertosaurus* and *Tarbosaurus* was profiled in a popular science book with the provocative title of *Dinosaur Gangs*, by Josh Young (Collins, 2011).

There has been a flood of studies using CT scans to study dinosaur brains. There are a couple of great reviews of the subject—how-to guides if you will—written by Carlson et al. (*Geological Society of London Special Publication*, 2003, 215: 7–22) and Larry Witmer and colleagues (in *Anatomical Imaging: Towards a New Morphology*, Springer-Verlag, 2008). The most important CT studies of tyrannosaurs are papers by Chris Brochu (*Journal of Vertebrate Paleontology*, 2000, 20: 1–6), by Witmer and Ryan Ridgely (*Anatomical Record*, 2009, 292: 1266–96), and by the Amy Balanoff and Gabe Bever duo and a team of collaborators (of which I am one) in *PLoS ONE* 6 (2011): e23393 and *Bulletin of the American Museum of Natural History*, 2013, 376: 1–72. Ian Butler and I published our first project on tyrannosaur brain evolution as part of our description of the new tyrannosaur *Timurlengia*, discussed in the previous chapter. Darla Zelenitsky's study of olfactory-bulb evolution was published in 2009 (*Proceedings of the Royal Society of London Series B*, 276: 667–73). Kent Stevens has published on binocular vision in tyrannosaurs (*Journal of Vertebrate Paleontology*, 2003, 26: 321–30).

Some of the most exciting recent work on tyrannosaurs—and dinosaurs more generally—uses bone histology to understand how they grew. I highly recommend two very readable reviews on the subject: a short paper written by Greg Erickson (*Trends in Ecology and Evolution*, 2005, 20: 677–84) and the book-length treatment by Anusuya Chinsamy-Turan (*The Microstructure of Dinosaur Bone*, Johns Hopkins University Press, 2005). Greg's landmark paper on tyrannosaur growth was published in *Nature* in 2004 (430: 772–75). Another important study on the topic was presented by Jack Horner and Kevin Padian (*Proceedings of the Royal Society of London Series B*, 2004, 271: 1875–80), and more recently the brilliant polymath Nathan Myhrvold (PhD in physics, former chief technology officer at Microsoft, frequent inventor, noted chef and author of the acclaimed *Modernist Cuisine*, plus a dinosaur paleontologist in his spare time) wrote an illuminating paper on the use, and sometimes misuse, of statistical techniques for calculating dinosaur growth rates (*PLoS ONE*, 2013, 8, no. 12: e81917).

Thomas Carr has written many papers on how *T. rex* and other tyrannosaurs changed as they grew. His most important works were published in the *Journal of Vertebrate Paleontology* (1999, 19: 497–520) and *Zoological Journal of the Linnean Society* (2004, 142: 479–523).

CHAPTER 7: DINOSAURS AT THE TOP OF THEIR GAME

I admit that my characterization of the latest Cretaceous as the apogee of dinosaur success is a bit subjective, and some of my colleagues may quibble with some of my statements. It comes down to the difficulty of measuring diversity in the fossil record, which is always subject to various biases, many of which we don't understand. There have been many studies of dinosaur diversity, including some that use statistical methods to estimate the total number of dinosaurs over time. These don't always agree in detail, but do agree on one general point: the latest Cretaceous was a time of generally high dinosaur diversity in terms of the number of recorded and/or estimated species. Even if it wasn't the absolute height of dinosaur diversity, it probably wasn't far off. My colleagues and I used different statistical methods to compute dinosaur diversity over the Cretaceous (Brusatte et al., *Biological Reviews*, 2015, 90: 628–42), finding that latest Cretaceous dinosaurs were either at or very close to their Cretaceous peak in species richness. Other important studies of dinosaur diversity over time have been published by Barrett et al. (*Proceedings of the Royal Society of London Series B*, 2009, 276: 2667–74); Upchurch et al. (*Geological Society of London Special Publication*, 2011, 358: 209–240); Wang and Dodson (*Proceedings of the National Academy of Sciences USA*, 2006, 103: 601–5), and Starrfelt and Liow (*Philosophical Transactions of the Royal Society of London Series B*, 2016, 371: 20150219).

Information on the history of the Burpee Museum can be found on the museum's website, http://www.burpee.org. Jane—the juvenile *T. rex* discovered by the Burpee Museum—is currently under study by a team led by Thomas Carr. A full description has not yet been published, but the fossil has been the subject of many Society of Vertebrate Paleontology conference presentations.

There is a wealth of information on the Hell Creek Formation. A good accessible primer is a review paper by David Fastovsky and Antoine Bercovici (*Cretaceous Research*, 2016, 57: 368–90). If you're looking for more detail, the Geological Society of America has published two special volumes on the Hell Creek (Hartman et al., 2002, 361: 1–520; and Wilson et al., 2014, 503: 1–392). Lowell Dingus has also written a popular book on the Hell Creek and its dinosaurs (*Hell Creek, Montana: America's Key to the Prehistoric Past*, St. Martin's Press, 2004). There have been two important surveys of Hell Creek dinosaurs, which is where I get the percentages of different species in the ecosystem. The first was led by Peter Sheehan and Fastovsky and published in a series of papers, including two particularly important works (Sheehan et al., *Science*, 1991, 254: 835–39; and White et al., *Palaios*, 1998, 13: 41–51). The second survey was conducted more recently, by Jack Horner and colleagues (Horner et al., *PLoS ONE*, 2011, 6, no. 2: e16574).

One of the best sources of information on *Triceratops*, and ceratopsians in general, is Peter Dodson's semitechnical book *The Horned Dinosaurs* (Princeton University Press, 1996). A more technical overview of these animals can be found in Dodson's chapter (cowritten with Cathy Forster and Scott Sampson) in *The Dinosauria* (University of California Press, 2004). Similarly, a prime source of information on the duck-billed hadrosaurs is the chapter by Horner, David Weishampel, and Forster in *The Dinosauria*, along with a recent technical book that includes several papers on the group (Eberth and Evans, eds., *Hadrosaurs*, Indiana University Press, 2015). There is also a chapter on the dome-headed pachycephalosaurs in *The Dinosauria*, written by Teresa Maryańska and colleagues, that is a good introduction to this bizarre group.

I was part of the team that described the Homer discovery—the first *Triceratops* bone bed—in the scientific literature. The paper was led by Josh Mathews, one of my fellow student volunteers on the 2005 expedition, and also included Mike Henderson and Scott Williams as coauthors (*Journal of Vertebrate Paleontology*, 2009, 29: 286–90). In this paper, we discuss and cite some of the other ceratopsian bone beds that had been found previously. A good review of ceratopsian bone beds, with citations to many important papers, was written by David Eberth (*Canadian Journal of Earth Sciences*, 2015, 52: 655–81). The *Centrosaurus* bone bed itself was described in a chapter coauthored by Eberth in the book *New Perspectives on Horned Dinosaurs* (Indiana University Press, 2007).

The best general reference on the dinosaurs of Late Cretaceous South America (and the southern continents more widely) is Fernando Novas's book *The Age of Dinosaurs in South America* (Indiana University Press, 2009). Roberto Candeiro has written many specialist papers on Brazilian dinosaurs, and some of his more important works on theropod teeth are his 2007 PhD thesis (Universidade Federal do Rio de Janeiro) and a 2012 paper (Candeiro et al., *Revista Brasileira de Geociências* 42: 323–30). Roberto, Felipe, and colleagues described a jawbone of a carcharodontosaurid from Brazil (Azevedo et al., *Cretaceous Research*, 2013, 40: 1–12), and Felipe's paper describing *Austroposeidon* was published in 2016 (Bandeira et al., *PLoS ONE* 11, no. 10: e0163373). The bizarre crocs of Brazil have been described in a series of publications (Carvalho and Bertini, *Geologia Colombiana*, 1999, 24: 83–105; Carvalho et al., *Gondwana Research*, 2005, 8: 11–30; and Marinho et al., *Journal of South American Earth Sciences*, 2009, 27: 36–41).

For some inconceivable reason, Baron Franz Nopcsa has yet to be the subject of a major biography or a film. There have been a handful of articles on him, however. The best of these are Vanessa Veselka's piece in the July–August 2016 issue of *Smithsonian*, an article by Stephanie Pain in *New Scientist* (April 2–8, 2005), and one by Gareth Dyke in *Scientific American* (October 2011). The paleontologist David Weishampel—who has spent many years excavating dinosaurs in Romania on the trail of the baron—has written often about Nopcsa. He paints an evocative picture of the baron in his 2011 book *Transylvanian Dinosaurs* (Johns Hopkins University Press) and also collaborated with Oliver Kerscher to collate a series of Nopcsa's letters and publications, which also includes a short biography and background to his scientific work (*Historical Biology* 25: 391–544).

Weishampel's book *Transylvanian Dinosaurs* is also the single best general reference on the Transylvanian dwarfed dinosaurs. For a more technical overview, there is a series of papers edited by Zoltán Csiki-Sava and Michael Benton, published as a special issue of *Palaeogeography, Palaeoclimatology, Palaeoecology* in 2010 (vol. 293). Helpful review papers have also been written by Weishampel and colleagues (*National Geographic Research*, 1991, 7: 196–215) and Dan Grigorescu (*Comptes Rendus Pale-*

ovol, 2003, 2: 97–101). I was part of a team led by Csiki-Sava who wrote a broader review of European latest Cretaceous faunas—there were actually several islands that dinosaurs lived on during this time, the Transylvanian one being the best studied and most famous (*ZooKeys*, 2015, 469: 1–161).

Mátyás Vremir, Zoltán Csiki-Sava, Mark Norell, and I published two papers on *Balaur bondoc*: a short initial description in which we named it (Csiki-Sava et al., *Proceedings of the National Academy of Sciences USA*, 2010, 107: 15357–61) and a longer monograph in which we figured and described each bone in detail (Brusatte et al., *Bulletin of the American Museum of Natural History*, 2013, 374: 1–100). With other colleagues, we also wrote a more expansive paper on the age and importance of the Transylvanian dinosaurs, with an emphasis on new discoveries (Csiki-Sava et al., *Cretaceous Research*, 2016, 57: 662–98).

CHAPTER 8: DINOSAURS TAKE FLIGHT

This chapter covers many of the themes that I wrote about in a *Scientific American* article (Jan. 2017, 316: 48–55), as well as in a technical review paper on early bird evolution (Brusatte, O'Connor, and Jarvis, *Current Biology*, 2015, 25: R888–R898) and a commentary piece in *Science* (2017, 355: 792–94). Much of the impetus for this chapter came from my PhD work, on the genealogy of birds and their closest relatives and on patterns and rates of evolution across the dinosaur-bird transition. I defended my PhD in 2012 (*The Phylogeny of Basal Coelurosaurian Theropods and Large-Scale Patterns of Morphological Evolution During the Dinosaur-Bird Transition*, Columbia University, New York) and published it in 2014 (Brusatte et al., *Current Biology*, 2014, 24: 2386–92).

There is a huge literature on the origin of birds and their relationships with dinosaurs. The best general readable sources of information are three review papers, written by Kevin Padian and Luis Chiappe (*Biological Reviews*, 1998, 73: 1–42), Mark Norell and Xu Xing (*Annual Review of Earth and Planetary Sciences*, 2005, 33: 277–99), and Xu Xing and colleagues (*Science*, 2014, 346: 1253293). Mark Norell's book *Unearthing the Dragon* (Pi Press, New York, 2005) is one of my all-time favorites—a romping journey through China to study feathered dinosaurs, enlivened with the photography of one of the best artists in the dinosaur business, my buddy Mick Ellison. More recently, Luis Chiappe and Meng Qingjin's *Birds of Stone* (Johns Hopkins University Press, 2016) is a beautiful atlas of feathered dinosaurs and primitive birds from China.

Pat Shipman's *Taking Wing* (Trafalgar Square, 1998) tells the tale of how scientists first recognized the dinosaur-bird link and the sometimes fierce debates as this once controversial hypothesis became mainstream. Huxley, Darwin, Ostrom, and Bakker are all covered here. Huxley laid out his theory on the dinosaur-bird link in a series of papers, including important ones published in the *Annals and Magazine of Natural History* (1868, 2: 66–75) and *Quarterly Journal of the Geological Society* (1870, 26: 12–31). Debates about *Archaeopteryx* are chronicled in Paul Chambers's book *Bones of Contention* (John Murray, 2002), which cites most of the relevant literature up to the early 2000s; the description of a new specimen of *Archaeopteryx* by Christian Foth and colleagues has recently taken the field much further (*Nature*, 2014, 511: 79–82). It is one of the papers that has advocated for a display "billboard" origin of theropod wings. The "Danish artist" is Gerhard Heilmann, and he made his arguments in his book *The Origin of Birds* (Witherby, 1926).

Robert Bakker has written the story of the Dinosaur Renaissance in a way that only he can, both in his *Scientific American* article (1975, 232: 58–79) and his book *The Dinosaur Heresies* (William Morrow, 1986). John Ostrom published a litany of careful scientific papers on the dinosaur-bird link, most important his meticulous monographic description of *Deinonychus* (*Bulletin of the Peabody Museum of Natural History*, 1969, 30: 1–165), his essay in *Nature* (1973, 242: 136), his review paper in *Annual Review of Earth and Planetary Sciences* (1975, 3: 55–77), and his masterful manifesto in *Biological Journal of the Linnean Society* (1976, 8: 91–182). It is also essential here to note that the pioneering cladistic analyses by Jacques Gauthier in the 1980s firmly placed birds among the theropods (e.g., in *Memoirs of the California Academy of Sciences*, 1986, 8: 1–55).

The first feathered dinosaur—*Sinosauropteryx*—was initially described by Qiang Ji and Shu'an Ji as a primitive bird (*Chinese Geology*, 1996, 10: 30–33). It was then reinterpreted as a feathered non-bird dinosaur by Pei-ji Chen et al. (*Nature*, 1998, 391: 147–52) and later described in detail by Phil Currie (Currie and Chen, *Canadian Journal of Earth Sciences*, 2001, 38: 705–27). Soon after the realization that *Sinosauropteryx* was a feathered dinosaur, an international team announced two additional feathered dinosaurs from China (Ji et al., *Nature*, 1998, 393: 753–61), and the floodgates opened from there. The vast majority of feathered dinosaurs discovered over the past two decades have been described by Xu Xing and his colleagues, and are well summarized in Norell's *Unearthing the Dragon*, as well as more recent literature cited in the review papers listed above. The preservation of the feathered dinosaurs, and the role of volcanoes in fossilizing them, has been studied by many authors, most recently and comprehensively by Christopher Rogers and colleagues (*Palaeogeography, Palaeoclimatology, Palaeoecology*, 2015, 427: 89–99).

The assembly of the bird body plan has been discussed by many authors. I've written about it in my PhD thesis and the *Current Biology* paper stemming from it (see above). Pete Makovicky and Lindsay Zanno covered it in their very readable chapter in the book *Living Dinosaurs* (Wiley, 2011). The American Museum Gobi expeditions are chronicled in one of my favorite pop-science dinosaur books: *Dinosaurs of the Flaming Cliffs* (Anchor, 1996) by Mark Norell's New York colleague, expedition coleader, and fellow SoCal surfer dude, Mike Novacek. Some of the more important research papers on the Gobi fossils—illustrating their importance in understanding the assembly of modern bird biology—have been the description by Norell et al. of the brooding oviraptorosaur (*Nature*, 1995, 378: 774–76) and the study by Balanoff et al. on the evolution of the bird brain (*Nature*, 2013, 501: 93–96). Background references on flow-through lungs and dinosaur growth have been outlined above, in the bibliography for previous chapters. The spectacular fossil of a dinosaur from Liaoning preserved in a birdlike sleeping posture was described by Xu and Norell (*Nature*, 2004, 431: 838–41), and the birdlike eggshell tissue was first identified in a dinosaur by Mary Schweitzer and colleagues (*Science*, 2005, 308, no. 5727: 1456–60).

Dinosaur feather evolution has been the subject of a great amount of research and an expansive literature. The review of

Xu Xing and Yu Guo (*Vertebrata PalAsiatica*, 2009, 47: 311–29) is a good starting point. For a developmental biology perspective on feather evolution, the many excellent papers of Richard Prum should be consulted. Darla Zelenitsky and her colleagues described their feathered ornithomimosaurs in 2012 (*Science*, 338: 510–14), and I gleaned details of their fieldwork from an October 25, 2012, article in the *Calgary Herald*. Jakob Vinther first presented his methodology for determining the colors of fossil feathers in a 2008 paper (*Biology Letters* 4: 522–25), which unleashed a number of studies by Vinther and others on feathered dinosaurs. All of this excitement is reviewed by Jakob in a review paper in *BioEssays* (2015, 37: 643–56) and a first-person piece in *Scientific American* (Mar. 2017, 316: 50–57). The flamboyant colors of early winged dinosaurs have been worked out by a Chinese-led team (Li et al., *Nature*, 2014, 507: 350–53), and the display function of wings has been discussed in a *Science* perspective piece by Marie-Claire Koschowitz and collaborators (2014, 346: 416–18). The wacky *Yi qi* was described by Xu and his team (*Nature*, 2015, 521: 70–73).

There has been an enormous—and often complex—literature on the flight abilities of early birds and feathered dinosaurs. A recent study by Alex Dececchi and colleagues—who found that *Microraptor* and *Anchiornis* were potentially capable of powered flight—is a good jumping-off point (*PeerJ*, 2016, 4: e2159). Engineering studies by Gareth Dyke and colleagues (*Nature Communications*, 2013, 4: 2489) and Dennis Evangelista and colleagues (*PeerJ*, 2014, 2: e632) deal with gliding in feathered theropods and review the most important previous work.

My colleagues and I presented our case for rapid rates of morphological evolution in early birds in a joint paper (*Current*

Biology, 2014, 24: 2386–92). The methods we used in that paper were developed with Graeme Lloyd and Steve Wang and described in an earlier work (Lloyd et al., *Evolution*, 2012, 66: 330–48). Roger Benson and Jonah Choiniere also demonstrated a burst of speciation and limb evolution around the dinosaur-bird transition (*Proceedings of the Royal Society Series B*, 2013, 280: 20131780), and Roger Benson's dinosaur body-size study (cited above) found the big decrease in size around this same point in the family tree. Many other recent studies have also looked at rates of evolution around the transition, and these are cited and discussed in the two above papers.

Jingmai O'Connor has named a copious number of new fossil birds from China. Two of her most important works are her genealogy of early birds (O'Connor and Zhonghe Zhou, *Journal of Systematic Palaeontology*, 2013, 11: 889–906) and her chapter (with Alyssa Bell and Luis Chiappe) in the book *Living Dinosaurs* (cited above). Her PhD advisor, Luis Chiappe, has also published a number of important papers on early birds over the last quarter century.

CHAPTER 9: DINOSAURS DIE OUT

I wrote about the dinosaur extinction for *Scientific American*, where I first told some of the stories in this chapter (Dec. 2015, 312: 54–59). After Richard Butler and I gathered our group of international colleagues to sit down and try to come to a consensus on the dinosaur extinction, we published a status report in *Biological Reviews* (2015, 90: 628–42). Joining Richard and me were Paul Barrett, Matt Carrano, David Evans, Graeme Lloyd, Phil Mannion, Mark Norell, Dan Peppe, Paul Upchurch, and

Tom Williamson. Additionally, Richard and I worked with Albert Prieto-Márquez and Mark Norell on our 2012 study of morphological disparity leading up to the extinction (*Nature Communications*, 3: 804).

My contribution to the dinosaur extinction debate, however, has been tiny. There have been hundreds, maybe thousands, of studies published on this greatest of dinosaur mysteries. There is no way I can do justice to all of them here, so instead I'll point inquisitive readers in the direction of Walter Alvarez's book *T. rex and the Crater of Doom* (Princeton University Press, 1997). It is a readable, entertaining, and scrupulous first-person view of how Walter and his colleagues solved the riddle of the end-Cretaceous extinction. It cites all of the most important papers on the subject, including those that lay out the evidence for impact, those that identified and dated the Chicxulub crater, and various dissenting views. The story that I tell at the beginning of the chapter, although full of artistic license, is based on the sequence of impact events Alvarez describes and the evidence that he outlines.

Much more work has been published since then, and much of this has been cited and discussed in our 2015 *Biological Reviews* paper. Some of the most exciting recent work—too recent to be discussed in our paper—is research by Paul Renne, Mark Richards, and their Berkeley colleagues that dates the Deccan Traps (the remnants of the big volcanoes in India), shows that most of the eruptions occurred right around the Cretaceous-Paleogene boundary, and argues that the asteroid impact may have kicked the volcanic system into overdrive (Renne et al., *Science*, 2015, 350: 76–78; and Richards et al., *Geological Society of America Bulletin*, 2015, 127: 1507–20). The timing of the Deccan eruptions and their relationship with the impact are still being debated as I write this.

Of course, anyone who's interested in the history of science and loves primary sources should check out the original paper in which the Alvarez team presented the asteroid theory (Luis Alvarez et al., *Science*, 1980, 208: 1095–1108), along with other papers from their team, and from Jan Smit and his colleagues around the same time.

Many independent studies have tracked dinosaur evolution during the Mesozoic, and many of them focus particularly on the latest Cretaceous. In addition to the new data set we presented in our *Biological Reviews* paper, the other key recent studies were published by Barrett et al. (*Proceedings of the Royal Society of London Series B*, 2009, 276: 2667–74) and Upchurch et al. (*Geological Society of London Special Publication*, 2011, 358: 209–40). Modern studies try to correct for sampling bias, but this is an issue that wasn't seriously recognized until a very important—but oddly, largely forgotten—paper by Dale Russell in 1984 (*Nature*, 307: 360–61). David Fastovsky, Peter Sheehan, and their colleagues grasped the lesson from this paper and published a very important study of latest Cretaceous dinosaur diversity in the mid-2000s (*Geology*, 2004, 32: 877–80). Jonathan Mitchell's ecological food-web study was presented in a 2012 paper (*Proceedings of the National Academy of Sciences USA*, 109: 18857–61).

The most important studies of Hell Creek dinosaurs and how they were changing in the lead-up to the asteroid impact include those by Peter Sheehan and David Fastovsky's team (*Science*, 1991, 254: 835–39; *Geology*, 2000, 28: 523–26), Tyler Lyson and his colleagues (*Biology Letters*, 2011, 7: 925–28), and the meticulous fossil catalogs of Dean Pearson and his collaborators, who include Kirk Johnson and the late Doug Nichols (*Geology*, 2001, 29: 39–42; and *Geological Society of America Special Papers*, 2002, 361: 145–67).

Fastovsky's undergraduate textbook, the subject of my praise in passing, is the excellent *Evolution and Extinction of the Dinosaurs* (Cambridge University Press), which was coauthored with David Weishampel. The book has gone through several editions, and is also available in a shorter and punchier version for younger students, called *Dinosaurs: A Concise Natural History*.

Bernat Vila and Albert Sellés have written many papers on the latest Cretaceous dinosaurs of the Pyrenees. The most general of these is a study of how dinosaur diversity changed in this region during the latest Cretaceous, a project that they generously invited me to contribute to (Vila, Sellés, and Brusatte, *Cretaceous Research*, 2016, 57: 552–64). Other important papers include Vila et al., *PLoS ONE*, 2013, 8, no. 9: e72579, and Riera et al., *Palaeogeography, Palaeoclimatology, Palaeoecology*, 283: 160–71. When it comes to Romania, the end-Cretaceous story is covered in the papers cited for chapter 7 above. Finally, Roberto Candeiro, Felipe Simbras, and I have written a paper summarizing the latest Cretaceous dinosaurs of Brazil (*Annals of the Brazilian Academy of Sciences* 2017, 89: 1465–85).

The question of why non-bird dinosaurs died, while other animals survived remains an active subject of debate. To my mind, the most important insights have been articulated by Peter Sheehan and his colleagues on the subject of plant- versus detritus-based food chains and land versus freshwater environments (e.g., *Geology*, 1986, 14: 868–70; and *Geology*, 1992, 20: 556–60); by Derek Larson, Caleb Brown, and David Evans on the subject of seed-eating (*Current Biology*, 2016, 26: 1325–33); by Greg Erickson and his team on the issue of egg incubation and growth (*Proceedings of the National Academy of Sciences USA*, 2017, 114: 540–45); and by Greg Wilson and his mentor Bill Clemens regarding mammal survivorship and the importance of small body size and generalist diets (e.g., Wilson's papers in *Journal of Mammalian Evolution*, 2005, 12: 53–76; and *Paleobiology*, 2013, 39: 429–69). An important paper by Norman MacLeod and colleagues is a good review of what lived and what died at the end of the Cretaceous, and what that may mean for kill mechanisms (*Journal of the Geological Society of London*, 1997, 154: 265–92).

I love the analogy of dinosaurs holding a "dead man's hand." I wish I could say I came up with it myself, but it was Greg Erickson who (as far as I know) first used it, in a quote in Carolyn Gramling's news article on his egg incubation study ("Dinosaur Babies Took a Long Time to Break Out of Their Shells," *Science* online, News, Jan. 2, 2017).

One further important caveat is necessary. The dinosaur extinction is probably the most controversial subject in the history of dinosaur research—at least judging by the number of hypotheses, research papers, debates, and arguments. The scenario that I present in this chapter—that the extinction happened suddenly and was primarily caused by the asteroid—stems from my own deep reading on the subject, my own primary research on latest Cretaceous dinosaurs, and in particular from the big communal consensus that we outlined in our *Biological Reviews* paper. I firmly believe that this scenario is most consistent with the evidence that we have, both in terms of the geological record (the evidence for a catastrophic impact is undeniable) and the fossil record (studies showing that dinosaurs were still quite diverse right up until the end).

There are, however, those with alternative views. The point of this chapter is not to dissect each and every theory on the dinosaur extinction—that could easily be the subject of an entire book—but it is worth giving some directions to literature that argues against my version of the extinction. For many decades, David Archibald and William Clemens have argued for a more gradual extinction, caused by changes in temperature and/or sea level; Gerta Keller and her colleagues have argued that the Deccan eruptions were the primary culprit; and more recently, my friend Manabu Sakamoto has used complex statistical models to make the iconoclastic claim that dinosaurs were in a long-term decline, in which they were producing fewer and fewer species over time. Dive into this literature to learn more, and decide for yourself where the preponderance of evidence lies. There are other skeptical or dissenting views as well, but that's all I'll have to say about that.

EPILOGUE: AFTER THE DINOSAURS

I told a small sliver of the New Mexico story in my *Scientific American* article on the rise of mammals (June 2016, 313: 28–35), coauthored with Zhe-Xi Luo. Luo is one of the world's experts on the early evolution of mammals. More important, he's a very generous and lovely guy. Like Walter Alvarez, Luo was on the receiving end of one of my brazen teenage requests. In the spring of 1999, when I was just turning fifteen, my family and I were set to take an Easter vacation to the Pittsburgh area. I wanted to visit the Carnegie Museum of Natural History, but not content with seeing only the exhibits, I desperately wanted a behind-the-scenes tour. I had read about Luo's discoveries of early mammals in the newspaper, then saw his contact details on the museum's website, so I got in touch. For an hour, he led my family and me on a tour of the bowels of the museum storehouse, and he still asks about my parents and brothers every time I see him.

My dear friend, colleague, and mentor Tom Williamson has made a career out of studying the Paleocene mammals of New Mexico, as well as the early evolution of placental mammals more generally. His magnum opus—which resulted from his PhD work—is his 1996 monograph on the anatomy, ages, and evolution of Paleocene mammals from New Mexico (*Bulletin of the New Mexico Museum of Natural History and Science*, 8: 1–141). Over the last few years, Tom has been leading me deeper into the dark side of mammalian paleontology. We've done joint fieldwork since 2011 and have started to publish some papers together, including a genealogy of primitive marsupials (Williamson et al., *Journal of Systematic Palaeontology*, 2012, 10: 625–51) and the description of a new species of beaver-size plant-eating mammal called *Kimbetopsalis* (the Primeval Beaver, as we cheekily call it), which lived just a few hundred thousand years after the dinosaurs died (Williamson et al., *Zoological Journal of the Linnean Society*, 2016, 177: 183–208). Tom and I currently co-supervise a PhD student who works on the Cretaceous-Paleogene extinction and the rise of mammals afterward: Sarah Shelley. Look out for her.

INDEX

Page numbers in *italics* refer to images.